T0205557

Studies in Computational Intelligence

Volume 765

Series editor

Janusz Kacprzyk, Polish Academy of Sciences, Warsaw, Poland
e-mail: kacprzyk@ibspan.waw.pl

The series "Studies in Computational Intelligence" (SCI) publishes new developments and advances in the various areas of computational intelligence—quickly and with a high quality. The intent is to cover the theory, applications, and design methods of computational intelligence, as embedded in the fields of engineering, computer science, physics and life sciences, as well as the methodologies behind them. The series contains monographs, lecture notes and edited volumes in computational intelligence spanning the areas of neural networks, connectionist systems, genetic algorithms, evolutionary computation, artificial intelligence, cellular automata, self-organizing systems, soft computing, fuzzy systems, and hybrid intelligent systems. Of particular value to both the contributors and the readership are the short publication timeframe and the world-wide distribution, which enable both wide and rapid dissemination of research output.

More information about this series at http://www.springer.com/series/7092

Alexander Gelbukh · Hiram Calvo

Automatic Syntactic Analysis Based on Selectional Preferences

 Springer

Alexander Gelbukh
Centro de Investigación en Computación
  (CIC)
Instituto Politécnico Nacional (IPN)
Mexico City
Mexico

Hiram Calvo
Centro de Investigación en Computación
  (CIC)
Instituto Politécnico Nacional (IPN)
Mexico City
Mexico

ISSN 1860-949X   ISSN 1860-9503   (electronic)
Studies in Computational Intelligence
ISBN 978-3-030-08908-5   ISBN 978-3-319-74054-6   (eBook)
https://doi.org/10.1007/978-3-319-74054-6

Printed on acid-free paper

This Springer imprint is published by Springer Nature
The registered company is Springer International Publishing AG
The registered company address is: Gewerbestrasse 11, 6330 Cham, Switzerland

Contents

Chapter 1
Introduction

1.1 Purpose

The most valuable treasure of humankind is knowledge. Computers have a better capability than humans to handle great amounts of information: search for information, apply simple inference, and look for answers to questions.... However, our treasure, which exists in the form of natural language texts—news boards, newspapers, and books in digital libraries and the Internet—is not understandable to computers; they deal with it as a chain of letters and not as knowledge.

This book describes several sentence analysis methods that are based on the syntactic and semantic characteristics of the sentences. It also describes a disambiguation algorithm that is based on linguistic and semantic knowledge obtained from a large quantity of text.

The methods described in this book pay special attention to correctly dividing structures that correspond to entities mentioned in a sentence; for example, {*the man with a gray coat who is standing in that corner*} is my father. Doing so enables us to perform further tasks, such as question answering, information retrieval, automatic translation, and logical text formalization.

We use several techniques to parse a sentence. We begin with simple heuristics and combine them to analyze more complex objects, such as subordinate and relative clauses. For example, a noun is most often preceded by an article (*the book, the house,* etc.), but then modifiers are added to that particular noun (i.e., object), such as adjectives or other descriptors. During this process, several ambiguities may arise; to solve these, we use selectional preferences.

Selectional preferences are conventionally determined based on the association between a pair of words. However, in several cases, selectional preference depends on several arguments, and not just one argument. For example, in *the cow eats grass* and *the girl eats ice cream*, it is unlikely that a girl eats grass and vice versa; therefore, the preference of *eat* depends on who is eating. A more complex situation can be found where more than one argument is involved to determine the preference

© Springer International Publishing AG 2018
A. Gelbukh and H. Calvo, *Automatic Syntactic Analysis Based on Selectional Preferences*, Studies in Computational Intelligence 765,
https://doi.org/10.1007/978-3-319-74054-6_1

of a third argument. Several proposals for handling dependency correlations in a sentence will be proposed over the course of this book. We will also describe new contributions for automatic selectional preference extraction as well as its many applications, and conventions for representing a sentence as a dependency structure will also be established.

1.2 Structure

We are mainly motivated in computational linguistics by the idea of an intelligent machine that is capable of understanding natural language and of conversing with human beings as if it were, itself, human. This ideal machine, as in science fiction movies, is still far from being realized. However, there are currently many useful tools for handling natural language. Advancing these tools is another goal of our study.

We begin with a rough description of **computational linguistics**, describing what it is and what its most common problems are, before focusing on sentence analysis. Our goal is outlined in Sect. 1.1, but there are several approaches by which it may be achieved. We explore several of these approaches in Chaps. 2 and 3, examining each approach's advantages and disadvantages. Chapter 2 examines sentence analysis of user utterances by **rewriting rules**. Chapter 3 looks at text representation through the **formalism of constituents**. Chapter 4 discusses the **formalism of dependencies**.

After exploring these approaches, at the end of Chap. 4, we present **the state of the art** of automatic extraction of semantic roles and selectional preferences. Some questions will remain, which we try to answer in the following chapters.

In Chap. 5, we describe our **evaluation** scheme and the results we obtained. First, we manually convert a syntactic annotated text to its dependency structure; then, we compare manual labels to automatically obtained labels.

In Chap. 6, we discuss applications of the parser outlined in Chap. 4, such as using it for word sense disambiguation (Sect. 6.1.4, Steganography 6.2). In Chap. 7, we propose solutions to the problem of **prepositional phrase attachment**.

In Chap. 8, we study an **unsupervised approach** with a "grammar induction parser."

Finally, in Chap. 9, we deal with the problem of multi-argument dependency, and in Chap. 10, we deal with the problem of needing all arguments of a verb to have full co-occurrence in order to predict them.

1.3 Field of Study

1.3.1 Natural Language and Computational Linguistics[1]

Linguistics studies human language. Within this vast science are sub-fields representing its intersection with other fields of scientific knowledge—for example, psycholinguistics or sociolinguistics—through technology, education, medicine, art, and other human activities. In particular, a special intersection exists between linguistics and computing that offers mutual benefits to both.

On the one hand, linguistics knowledge is the theoretical basis for the study of a large range of highly important technological applications for our incipient information society—for example, the searching and handling of knowledge, natural language interfaces between humans and computers or robots, and automatic translation.

On the other hand, computational technologies offer tools to linguists. Until two decades ago, researchers could not have even dreamed of these tools and, until recently, linguists could not make everyday use of them due to the prohibitive cost of computing power and/or data storage. To name a few, these tools include immediate search results for any given word's usage and construction within large amounts of text; complex statistics obtained incredibly fast; the almost-instantaneous (compared with the speed of pencil and eraser) analysis, marking, and classification of any text; and automatic detection of the structure of an unknown language. The advanced search engines of the Internet have opened the door to a whole world of language—to a corpus so vast, it could reasonably be stated that all human language is available in a palpable and measurable way. In comparison, a conventional corpus represents a mere drop in the vast ocean of humanity's collective use of language.

Among these benefits is the possibility of the massive verification of theories, grammars, and linguistic dictionaries, which is simply outstanding. Some years ago, to verify a colleague's proposed grammar, a linguist had to use his or her intuition in searching for examples not covered by the proposed grammar; if no such example was found, the linguist had to suppose the grammar complete. Nowadays, automatic grammar analyzers will not only verify whether a grammar is complete, they will also quantifiably measure to what extent the grammar is complete and the productivity of each of the grammar's rules.

However, the main benefit that computer technologies offer linguistics in all its subfields (from lexicography to semantics and pragmatics) is the motivation to compile a language description in a complete and precise way—that is, formally.

More specifically, this relationship can be described in the following way. Linguistics, as every science, constructs the models and descriptions of its object of study: natural language. Conventionally, such descriptions were focused on the human reader, in many cases resorting—sometimes even without the author's notice—to the reader's common sense, intuition, and own knowledge of language.

---

[1]Based on [85].

Historically, the first challenge for such descriptions was the description of any given foreign language, where it was impossible to resort to the reader's own linguistics sense; such descriptions of foreign languages actually helped greatly improve the clarity of language descriptions, now called "formality." However, even in these descriptions, analogies with the reader's own language were often implicitly used for support, not to mention persistent references to common sense.

The computational revolution has given linguists an interlocutor with a singular property: without having either prior knowledge, intuition, or common sense, computers are highly capable of interpreting and literally applying the language descriptions provided by linguists. Just as we struggle to answer children's questions about things we once thought obvious, computers force linguists to sharpen and complete their formulations. In this way, computing assists linguistics (a traditionally humanistic science) to express terms formally.

The wide field where linguistics and computing interact and intersect is, in turn, structured into several (more specific) sciences. In particular, the science of computational linguistics is about the construction of language models that are *understandable* to computers—that is, more formal than conventional models, which are oriented to human readers.

1.3.2   Levels of Linguistic Processing[2]

From a technical point of view, natural language processing (NLP) aims to practically solve some issues that are studied by computational linguistics. NLP is highly complex because of the amount of knowledge involved. Compilation of this knowledge is one problem of linguistic systems engineering, and automatic learning from large-scale text corpora is a solution commonly used to solve that problem.

Another solution is to partition processing into steps (phases) that correspond to the levels (layers) of language: morphological (word) analysis, syntactic (sentence) analysis, and semantic (whole text) analysis. Yet, this solution yields another problem: ambiguity. Ambiguities produced in one level (for example, is *test* a noun or verb?) must be solved in other levels of analysis. Ambiguity is probably the most important problem in natural language analysis. Traditionally, natural language has been divided into the following *levels of language*:

1. phonetics/phonology
2. morphology
3. syntax
4. semantics
5. pragmatics
6. discourse

---

[2]Based on [85].

There are no particular criteria for separating each level; rather, the differences between levels are based on the analysis approach used in each one. Therefore, there may be overlap between levels, which does not necessarily indicate contradiction. For example, some phenomena are related to both phonology and morphology (for example, stem alternations such as *mouse-mice* and *do-did*).

Each language level and their computational advances will be briefly discussed in the following subheadings.

1.3.2.1   Phonetics/Phonology

Phonetics is the area of linguistics devoted to the exploration of sound characteristics, which is a substantial form of language. Because phonetics is devoted to sound characteristics, its methods are mostly physical, which makes its position in linguistics quite independent.

Problems in computational phonetics are related to the development of voice recognition and speech synthesis. Even though there are voice recognition systems—that is, the computer can recognize words uttered in front of a microphone—the percentage of correctly identified words is relatively low. Among systems of speech synthesis, there is much more success. Some such systems are able to speak with a *robotic* accent but do not sound completely human (several generation modules are available for testing at loquendo.com). Furthermore, speech synthesis systems have a rather restricted area of application: usually, it is much faster, comfortable, and safe to read a message than listen to it. Thus, speech synthesis systems are basically only truly useful to people with sight deficiencies.

Phonology is also interested in sounds, but from another point of view. Phonology's interest lies in the position of a sound within the sound system of a certain language—that is, the relationships a sound has with the other sounds in the system and the implications of such relationships. For example, why can Japanese speakers not easily distinguish between the [r] and [l] phonemes? Why do foreigners speak Spanish with a noticeable accent—such as pronouncing [r] as [rr]? Why do native Spanish speakers usually have accents when speaking certain other languages —for example, not being able to pronounce [hard l] as *l* is pronounced in English? The answer is the same in all cases: their native languages have no opposition between the phonemes mentioned; because of this, differences that seem very noticeable in some languages are insignificant in others. In Japanese, there is no [l] phoneme; in the majority of languages, there is only one phoneme for [r]-[rr] and, obviously, its duration is not important (Spanish represents the contrary case). On the other hand, in Spanish there is no [hard l] phoneme—only [l] (soft l) exists— thus, when native Spanish speakers speak English, they pronounce [hard l] softly as in their mother tongue.

1.3.2.2   Morphology

Morphology looks at the internal structure of words (suffixes, prefixes, stems, inflections) and the system of grammatical language categories (gender, number, etc.). Some languages are vastly different from English. For example, in Arab, the stem or root has three consonants and the diverse grammatical forms of a word are made by means of inserting vowels between the consonants (*KiTaB* <*the book*>, *KaTiB* <*reading*>); in Mandarin, there is almost no morphological form for words, which is compensated for on the syntax level (by using a fixed word order, auxiliary words, etc.); in Turkish languages, each suffix attached to a root expresses a single value of a grammatical category. For example, in Azerbaijan, the single form *baj-dyr-abil-dy-my* (which has four grammatical morphemes) means "if he could make see;" these four morphemes express possibility (can), obligation (make), past, and interrogation. Thus, it is not possible to translate this word with a single English word because the morphemes (which are grammatical and inside the word in Azerbaijan) correspond to auxiliary verbs in English. (Note that it is possible to build words in Azerbaijan that have more than 10 morphemes.)

Problems in computational morphology are related to the development of systems of analysis and automatic morphological synthesis. The development of such modules still requires great effort because they are required in order to build great root dictionaries (whose entries should number in the hundred thousands). In general, there is a methodology for such development, and systems are currently working for many different languages. The problem, however, is to design a standard for such modules.

1.3.2.3   Syntax

The main task at the syntax level is to describe how words in a sentence are related and what function each word performs—that is, to build the structure of the sentence of a language.

Rules for building sentences are defined prescriptively for humans: correct forms are given and deviations from those forms are banned. In other words, the preferred usage for that language is clearly provided. In contrast, rules for the linguistic processing of texts must be descriptive, establishing methods to define both possible and impossible phrases for a specific language.

Possible phrases for a specific language are grammatical sequences, that is, they obey grammatical laws and do not require contextual knowledge; ungrammatical phrases must be propagated to levels that consider context and reasoning in a wider sense. Establishing methods that uniquely determine grammatical sequences during the linguistic processing of text has been the goal of grammar formalisms in computational linguistics. In that field of study, mainly two approaches have been considered: dependencies and constituents. In dependencies, relationships are marked with arrows and a word can have several words depending on it. In constituents, relationships are represented as a binary tree.

The study of computational syntax requires methods for both automatic analysis and synthesis—that is, methods for constructing the phrase structure and methods for generating the phrase based on its structure. Developing generators is an easier task, where it is clear which algorithms are necessary for these systems. On the contrary, developing syntactic analyzers (also called *parsers*) is still an open problem, particularly for languages that do not have a fixed word order, like Spanish. In English, the word order is fixed and, because of this, English-based theories are not so easily adapted to other languages. We will present a *parser* in the later sections.

1.3.2.4 Semantics

The purpose of semantics is to *understand* the phrase. But what does it mean *to understand*? In order to understand, it is necessary to know the meaning of all words and to interpret their syntactic relationships. Researchers partially agree that the results of semantic analysis should be semantic networks, where all concepts and their relationships are represented. Another possible representation is something very similar to semantic networks: *conceptual graphs*. Then, what we need to know is how to transform a syntactic tree into a semantic network. That problem still does not have a general solution.

Another task of semantics (or, more specifically, its subfields of lexicology and lexicography) is to define word senses. This is a difficult task, whether done manually or automatically. The results of such sense definitions can be seen in the form of dictionaries. The main problem here is that there is always[3] a vicious circle in a word's definition because all words are defined through other words. For example, defining *cock* as "the male of a *hen*" and *hen* as "the female of a *cock*" will not help someone wanting to learn what a *cock* and *hen* are. In this example, the vicious circle is very short; usually, circles are larger—but they are avoidable. Computational semantics can help by looking for a set of words through which other words are defined. The set of words is called the *defining vocabulary*. Computational semantics can also help in evaluating the quality of dictionaries—as we all know, there are both good and bad dictionaries available.

An important application of semantic analysis is *word sense disambiguation*. For example, a *bank* can be an institution, a place to sit, or a place where fish live. It is possible to determine which of these meanings is meant by analyzing other words used in context. For example, in *I went to the bank to withdraw money*, the words *withdraw* and *money* show that this use refers to the *institution*; meanwhile, *the bank is very clean and I can see the fish* refers to a bank where fish live. However, in *that bank is very good and efficient*, it is impossible to automatically determine the correct definition—without greater context, it cannot even be done manually.

---

[3]If there is no vicious circle, then some words are not defined [84].

In summary, a lot of research is still needed in the field of computational semantics.

1.3.2.5   Pragmatics

It is often said that pragmatics is about relationships between a sentence and the external world. A famous example is where you and I are eating together and I ask you if I can have the salt; you say, "yes," but keep eating. Surely, the response in that example is formally correct because I *can* have the salt (that is the literal meaning of the question), but my intention was to request the salt, and not to ask about the possibility of my having it. In other words, we can say that pragmatics studies the intentions of the author or speaker.

Pragmatics also studies sentences that are interesting in that they, themselves, are the action. This type of sentence is called a performative, and a good example is the sentence *I promise*. Note that *I promise* is precisely the action of *to promise*.

Because many difficulties are often found at the semantic level, it is usually difficult to continue analysis into the next level (discourse), but it is also a level worthy of consideration.

1.3.2.6   Discourse

Usually, we do not talk in isolated sentences. Rather, we most often communicate with a string of sentences, which have certain associations with one another, making them more than just sentences. Indeed, what we have then is a new entity called a *discourse*.

A very important problem exists in the analysis of discourse: *co-reference*. A certain type of *co-reference* is called *anaphoric reference*; it points to previous references.

For example, the discourses "I saw a new house yesterday. Its kitchen was exceptionally big" (its = house's) and "John arrived. He was tired" (he = John) contain co-reference relationships, which the computer needs to interpret in order to build semantic representations.

Very good algorithms for co-reference resolutions exist and are correct up to 90% of the time; however, solving the other 10% is difficult.

1.3.3   Ambiguities in Natural Language

Ambiguity in a linguistic process occurs when several interpretations can be admitted from the representation or when it is not possible to determine which structure is correct. To *disambiguate*—that is, to select the most adequate meaning or structure from a known set of possibilities—several solution strategies are required.

Related to syntax, there is ambiguity in the marking of parts of speech. This ambiguity refers to the fact that one word can have several syntactic categories; for example, *on* could be a preposition or an adverb, and *can* could be a verb or a noun (tin can). Knowing the correct category for each word in a sentence aids in syntactic disambiguation; however, disambiguation of this categorizing requires, in turn, a certain kind of syntactic analysis.

Syntactic analysis deals with several forms of ambiguity. The main ambiguity occurs when the syntactic information is not enough to make a decision of structural assignment. Such ambiguity exists even for native speakers—that is, there are different interpretations for a single phrase. For example, in *John sees a man with a telescope*, two different interpretations are possible: John uses a telescope to see a man or John sees a man who has a telescope.

Ambiguity also exists in predicative complements. For example, in *I like to drink wine with my friends*, the phrase *with my friends* is a complement of *to drink*—not of *wine*. A human speaker would not consider the possibility of grouping *wine* and *with my friends*, but both are real possibilities to a computer.

As mentioned, lexical information can be helpful when solving many ambiguities. In other cases, semantic proximity helps to disambiguate. For example, in both *I like to drink liquor with mint* and *I like to drink liquor with my friends*, the semantic class of the last noun helps solve the ambiguity—which part of the phrase is linked to the prepositional phrases (PPs) *with mint* and *with my friends?* Note that neither *mint* nor *friends* is ambiguous by itself; however, *friends* is semantically closer to *drink* than to *liquous*, and *mint* is semantically closer to *liquor* than to *drink*.

In this book, we focus mainly on syntactic analysis along with semantic features (mainly, lexical semantics) in order to simultaneously solve both word and structure ambiguities.

Chapter 2
First Approach: Sentence Analysis Using Rewriting Rules

Some of the earliest useful user interaction systems involving sentence analysis used rewriting rules. A famous example of one such system is SHRDLU, which was created in the 1960s by Terry Winograd at the Massachusetts Institute of Technology and was able to solve many of the problems that arise when conversing with a computer.

SHRDLU is not an acronym. The name is based on the frequency with which letters are used in the English language; in order, the most frequently used letters in English are E, T, A, O, I, N, S, H, R, D, L, and U. SHRDLU was designed with several rows of linotypes representing every letter in the English language; these rows were arranged from the most frequently used letters to the least frequently used ones—thus, the first row of linotypes spelled ETAOIN and the second row spelled SHRDLU. Winograd chose this name inspired by a science fiction story written by Fredric Brown[1] in 1942. This story is about an intelligent linotype that learns everything it types and then, eventually, attempts world domination.

SHRDLU's purpose was to understand natural language. It allowed users to dialogue with a computer (using a keyboard and screen) about a small selection of objects, which were shown with a cathode ray tube (CRT) screen connected to a PDP-6 computer. By restricting its domain to a very small selection of objects (a *micro-world*), the program was able to support dialogues wherein the computer seemed to understand anything requested by the user.

---

[1]http://www.sfsite.com/isfdb-bin/exact_author.cgi?Fredric_Brown

© Springer International Publishing AG 2018
A. Gelbukh and H. Calvo, *Automatic Syntactic Analysis Based on Selectional Preferences*, Studies in Computational Intelligence 765, https://doi.org/10.1007/978-3-319-74054-6_2

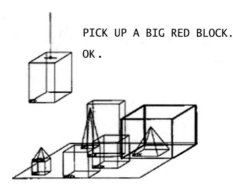

PICK UP A BIG RED BLOCK.

OK.

The blocks micro-world (image taken from http://hci.stanford.edu/∼winograd/shrdlu/)

Person: PICK UP A BIG RED BLOCK.
Computer: OK. (does it)
Person: GRASP THE PYRAMID.
Computer: I DON'T UNDERSTAND WHICH PYRAMID YOU MEAN.
Person: FIND A BLOCK WHICH IS TALLER THAN THE ONE YOU ARE HOLDING
AND PUT IT INTO THE BOX.
Computer: BY "IT," I ASSUME YOU MEAN THE BLOCK WHICH IS TALLER THAN
THE ONE I AM HOLDING.
Computer: OK. (does it)
Person: WHAT DOES THE BOX CONTAIN?
Computer: THE BLUE PYRAMID AND THE BLUE BLOCK.
Person: WHAT IS THE PYRAMID SUPPORTED BY?
Computer: THE BOX.
Person: HOW MANY BLOCKS ARE NOT IN THE BOX?
Computer: FOUR OF THEM.
Person: IS AT LEAST ONE OF THEM NARROWER THAN THE ONE WHICH I TOLD
YOU TO PICK UP?
Computer: YES, THE RED CUBE.
Person: IS IT SUPPORTED?
Computer: YES, BY THE TABLE.
Person: CAN THE TABLE PICK UP BLOCKS?
Computer: NO.
⋮

This dialogue shows very advanced features of natural language understanding: micro-world knowledge, understanding of question and request, reference resolution (*this, the one I told you to pick up*), etc. These characteristics allow the user to manipulate a computer using natural language.

In general, human-machine communication in real world situations is carried as dialogue consisting of a user's requests for information or action and the computer's verbal- or action-based response, In this chapter, we present a system that deals with this kind of interaction through a rewriting rules grammar; this system has certain features—such as modification of an object's properties, substitution using wildcards, and in-line functions. Our goal is to produce one or more specific computer instructions derived from the user's requests. Grammar rules and special context objects (called "scenes") translate objects referenced within the request into symbols. Then the instructions are executed by an external system. Examples are given where the computer was requested to place 3-dimensional objects following natural language instructions; the natural language used was Spanish.

Spatial object composition (SOC) refers to manipulating physical or virtual prefabricated pieces (such as furniture parts) in order to assemble them into new objects or scenes (such as office design). Many computer applications—such as computer-aided design programs—deal with SOC. Objects that are to be placed in the room are predefined (furniture, doors, windows, etc.) and can be selected from a catalog and set in the virtual scene the user selects.

Obviously, SOC is not limited to house design. Since we live in a spatial world of decomposable objects, many various applications exist. For example, suppose a user wants to construct a bookcase. To do so, he or she first selects planks and then fits those parts together until the desired bookcase is constructed. This is an example of creating a new object.

By their nature, such systems are intended to be used by people without any computer-related knowledge or skill. Thus, their interfaces must be intuitive and not require any training or instruction. Natural language is, of course, a perfect means of attaining such interaction and would allow users to interact with the computer similarly to how they would interact with human workers. Indeed, human-computer interaction in such systems is mostly imperative: the user gives a command and the computer executes the requested task. These commands can be given in natural language imperatives. This is the root of our motivation to develop a framework for integration of natural language interfaces with SOC systems.

Within the proposed framework, it is possible to translate the input sentence *Could you put the chair next to the table please?* into a sequence of commands directly interpretable by the system's engine: `move(obj_chair1,getpos (obj_table1)++)`, which means that the system must place `obj_chair1` (the chair the user referred to) as near as possible to the position of `obj_table1` (the table already existing in the scene).[2] We do this by transforming the original sentence step-by-step, as follows:

Could you put the chair next to the table, **please?**
Could you put the chair next to the table

---

[2]Here, `obj_` stands for an object, `getpos` for a function that gets the position of the object, and ++ for the operation that changes a position to the nearest available.

> put **the chair** next to the table
> put obj_chair1 next to **the table**
> put obj_chair1 **next to** obj_table1
> put obj_chair1 **nextto obj_table1**
> **put** obj_chair1 **(getpos(obj_table1)++)**
> move(obj_chair1,(getpos(obj_table1)++))

An action-request system is an integrated system in which the user interacts with a computer by posing questions or requesting actions to be performed within a particular area or subject. The computer answers or performs the requested action. This kind of system has a specific domain and a limited number of objects and actions that can be performed on them. This limitation allows us to think of a natural language interface where the requested actions and asked questions are posed to the system in a free and natural way. The system should react to these requests by completing (or answering) what was requested in a coherent way. In this way, the user can acknowledge that the system understood his or her request.

Our system will be based on a rewriting rules grammar; however, our goal is different from that of generative grammars, which are created to verify whether the sentences of a particular language are well formed [175]. Because our goal is different, we could say that we are not following the traditional concept of a generative grammar. Our formalism is a rewriting rules grammar but is different from other such grammars in that ours includes the ability to modify properties, perform substitution by means of wildcards, and allows the use of in-line functions. The purpose of this grammar is to reduce language expression to a logic form (a set of instructions) by applying the rewriting rules directly to users' expressions.

To better show how our system works, we assume a task where geometric objects—such as spheres, toruses, planes, and other geometric figures—can be combined on a three-dimensional canvas. This graphical task was chosen because it allows the representation of objects to be shared between humans and computers since it is easy to visualize. It is important to make the context for the user and system evident; without this particular feature, the search for referenced objects would not be possible, as explained later.

Additionally, a visual task like the one proposed allows us to create more complex objects through constructive solid geometry (CSG), using operators like union, intersection, difference, and combination (merge) [94]. In this way, new objects can be created. All objects can have properties (such as texture, reflectivity, refraction index, etc.) that can be modified through natural language.

As discussed in the introduction to this chapter, one of the first systems to handle objects through natural language was SHRDLU. Though our system might seem similar to SHRDLU, our goal is not to recreate it but, rather, to create a system that can handle both tasks that involve objects in a spatial environment and generic instructions in natural language.

An example of the kind of expression that users can pose to the system is *¿Me puedes poner el tercero junto al toroide?* (Can you put the third one next to the torus?). After applying seven rewriting rules, which we will describe in detail later (see Sect. 2.6), we obtain the following instructions:

`move(objcat021_03, getpos(obj002)); sos.push(obj003); oos.-push(obj002).`

The instruction `move(objcat021_03)` will be performed by an external function that moves the requested object. `sos.push(obj003)` and `oos.push(obj002)` are in-line functions used for handling the object and the context. These functions will be detailed in Sect. 2.4.2.6. In the following sections we will present related work, details of the grammar, its mechanism for object and context handling, and, finally, processing examples for four sentences.

2.1 Related Work

Historically, the first systems with natural language interfaces (NLIs) were developed on an ad hoc basis for specific applications. Some examples include:

- DEACON (Direct English Access Control), which allowed computers to answer questions [62];
- SHRDLU, which moved virtual geometric blocks using verbal commands from human users [201];
- LUNAR, which allowed human users to query a lunar rocks database [203]; and
- LADDER, which answered questions in the English language about naval logistics data [100].

Since the world model was interwoven into these programs' operations, changing their application domains would be an expensive and complicated process.

Later, other systems with NLIs, designed with a broader scope of application, were created. These were mainly oriented to database information retrieval—for example, INTELLECT [99], TEAM [97], JANUS [200], and SQUIRREL [9].

Some recently developed systems handle natural language imperatives for multiple purposes. For example, KAIRAI (which means "puppet") has several virtual robots (avatars) that can move forward, turn, or push an object [5, 181]. By manipulating these robots with commands, the user can move and place objects in the virtual world. (KAIRAI is developed for use only with the Japanese language.) A similar system, AnimAL, uses an natural language interface (NLI) to control the movements of an avatar in a virtual environment [66, 67, 197]. Di Eugenio, who helped design AnimAL [66, 67], treated the problem of understanding phrases such as *do x to do y* as *cut a square in half to make two triangles*.

We are not aware, however, of any recent works specifically devoted to providing an NLI framework to SOC systems in general.

2.2 Characteristics of SOC Systems

SOC systems, in general, restrict the use of natural language in a number of ways. In our framework, we rely on these restrictions to simplify the corresponding mechanisms. Specifically, SOC systems have the following characteristics relevant to designing an NLI:

1. **They have predefined basic objects that can be used to construct new ones**, which allows us to begin with a reduced set of object names that must be recognized.
2. **Objects have properties** by which they can be referred, e.g., *red plank* as opposed to *green plank*. Properties let us keep our set of object names small.
3. **There is a visual-spatial representation common to the user and computer**. With this, the user is aware that the only existing objects are those that can be observed in the catalogs and current scene. Only observable objects are relevant to the composition task.
4. **Objects have a limited number of actions that can be applied to them**. These can be mapped to corresponding computer instructions.

The user and computer manipulate a finite set of objects, which have properties and actions attached to them. To design a suitable NLI, we must find a mechanism that relates natural language sentences to corresponding computer instructions. This relation is implemented through the direct translation grammar presented in the next section.

2.3 Direct Translation Grammar

Since Chomsky's transformational model first appeared in 1957 [50], a number of models within the generative paradigm have been suggested, such as case grammars [76], functional grammars [106], and, recently, phrase structure grammars [83]. Traditionally, generative grammars are designed to model the whole set of sentences that a native speaker of a natural language considers acceptable [161]. Generative linguistics views language as a mathematical object and builds theories similar to the sets of axioms and inference rules found in mathematics. A sentence is grammatical if there is some derivation that demonstrates that its structure corresponds to the given set of rules, much as a proof demonstrates the correctness of a mathematical proposition [202].

Phrase structure grammars (PSGs), of which HPSG (Head-driven Phrase Structure Grammar) [175] is the most widely known, follow this generative paradigm. To analyze a sentence, that sentence is first hierarchically structured to form phrase-structure trees. PSGs are used to characterize these phrase-structure trees and consist of a set of non-terminal symbols (phrase-structure categories such as noun, verb, determiner, preposition, noun phrase, verbal phrase, sentence, etc.), a set of terminal symbols (lexical items such as *buy*, *John*, *eaten*, *in*, *the*, etc.), and a set of rules that relate a non-terminal symbol with a string of either terminal or non-terminal symbols [105]. To analyze a sentence, suitable rules can be applied to the terminal symbol string until the non-terminal symbol S is reached.

The phrase-structure tree obtained during this process can be analyzed later to generate computer instructions equivalent to the input sentence. However, this process can be done directly if we change the purpose of our grammar such that the grammar rules are used to directly reach computer instructions—instead of using the grammar rules to break down natural language sentences into parts of speech (phrase structures) and then converting the structure of those sentences to computer instructions. Thus, our focus is different from that of generative grammars in that we are not interested in determining if a sentence is well formed. In addition, we are not interested in modeling the whole language but only the small subset that is relevant to the user's current task.

The grammar we propose for translating natural language sentences into computer instructions is a rewriting rules grammar with additional characteristics included in order to handle context and object references. We call it the "direct translation grammar" (DTG).

Within DTG, lexical and morphological treatments are included, and the categories used refer to both syntactic and semantic concepts of the sentences. Therefore, we can consider DTG a semantic grammar [32]. In semantic grammars, the choice of categories is based on the semantics of the world and the intended application's domain as well as on regularities of the language. Although they are not currently widely used, semantic grammars have several advantages—such as efficiency, habitability (in the sense of Watt [196]), handling of discourse phenomena, and the fact that they are self-explanatory. Such grammars allow the use of semantic restrictions to reduce the number of alternative interpretations that can be considered at a certain moment, in contrast to highly modular systems, which fragment the interpretation process.

2.4 Definition

We define a direct translation grammar as an ordered list of rewriting rules that have the form $\alpha \rightarrow \beta$, where α and β are strings consisting of one or more of the following elements, in any order:

1. natural language words,
2. tags with properties,
3. wildcards,
4. names of external procedures,
5. symbolic references to objects, and
6. embedded functions for context control and object reference handling.

Two or more rules with the same α are not allowed.

2.4.1   Rule Order

Rule processing is ordered. The rules for α, consisting only of natural language words, are first considered—beginning with those rules that have a greater number of words than the others. If no rule can be applied, the rest of the rules are considered according to the number of elements that compose α, with rules containing the most being considered first. Rules with the most elements are considered first because elements such as "the red table" must be considered before elements such as "the table" can be. Indeed, a longer string of words means a more specific reference to any given object.

Each time a new rule is applied, rule processing restarts from the top of the list in the order just explained. The process finishes when no rule can be applied; the resulting string at that point is the output of the program. The translation process is considered successful if the resulting string consists only of symbolic references to objects and names of external procedures. To avoid infinite cycling, the process is aborted if some rule is applied more than once and its application results in a previously obtained string; in such a case, translation is considered unsuccessful.

2.4.2   Rule Components

In this section, we explain each element that is used in the rules listed in Sect. 2.4.

2.4.2.1   Natural Language Words

Initially, an input sentence consists only of words. *Put the chair next to the table* is a sentence composed of seven words; in the sub-sections that follow, that sentence will be translated into a sequence of external procedures. Words are letter strings and do not have any properties.

2.4.2.2   Tags with Properties

Tags with properties have the form

$$\delta\{p_1, p_2, \ldots, p_n\},$$

where δ is the name of the tag and p_1, p_2, ..., p_n are its properties in the form name:value, e.g., put{C:V, T:IMP}. In Table 2.1, we present the most common properties and their possible values.

Table 2.1 Common properties and the values used for them in the in-text examples

Name	Property	Possible values
C	Category	N (noun), V (verb), ADJ (adjective), ADV (adverb), PRO (pronoun), DEFART (definite article), INDART (indefinite article), OBJ (object), POS (position)
G	Gender	M (masculine), F (feminine), N (neutral)
N	Number	S (singular), P (plural)
T	Verbal tense	PRES (present), INF (infinitive), IMP (imperative), SUBJ (subjunctive)
S	Subject form	For verbs: the number and gender of the subject (this is morphologically relevant for Spanish)
O	Object form	For verbs: the number and gender of the object (morphologically relevant for Spanish)
A	Dative object form	For verbs: the number and gender of the indirect (dative) object (morphologically relevant for Spanish)
Q	Quantity	V, L, R, U, M (very little, little, regular, much/many, very much/many)

This construction resembles traditional feature structures. However, feature structures, as defined by Kay in [106], undergo inheritance mechanisms and unification. Our tags are not related to such mechanisms.

The following rule converts the Spanish word *pon* "put$_{\text{imperative}}$" into a tag:

$$\texttt{pon --> poner\{C:V, T:IMP, S:2S, A:1S\}.}$$

This rule substitutes every occurrence of *pon* in the input string by the tag `poner{C:V, T:IMP, S:2S, O:1S}` with the following properties: category being verb, tense being imperative, subject being of second person singular, (implicit) dative object being of first person singular.

2.4.2.3   Wildcards

Wildcards are defined by a label that may optionally be followed by a set of properties (as defined in Sect. 2.4.2.2) contained in square brackets:

$$\varphi[p_1, p_2, \ldots, p_n].$$

Wildcards allow rules to be generalized in order to avoid redundant rule repetitions; they also make it possible to apply a rule over a set of tags that share one or more properties. The scope of a wildcard is limited to its rule.

A wildcard φ matches a tag δ if δ has all the properties listed for φ at the same values. For example, both wildcards `A[C:V]` and `B[T:IMP, S2S]` match the tag `poner{C:V, T:IMP, S:2S, O:1S}`, but `C[C:V, T:PRES]` does not because tag `poner{C:V, T:IMP, S:2S, O:1S}` does not have the property *Tense* with value *Present*.

When used in the right-hand side of the rule, a wildcard can be used to modify properties by specifying another value for the property that it originally matched. For example, consider the collocation *podrías juntarlo* ("could you put it together") which is a polite euphemism for the imperative *júntalo* ("put it together"). To transform the collocation into an imperative, we first apply the following rules:

$$podrías - - > poder\{C:V, T:SUBJ, S:2S\} \qquad (2.1)$$

$$juntarlo - - > juntar\{C:V, T:INF, O:3SM\}. \qquad (2.2)$$

We then use a wildcard to transform any such construction into an imperative; note the use of a wildcard to change the property T from INF to IMP:

$$poder\{C:V, T:SUBJ, S:2S\}A[C:V, T:INF] - - > A[T:IMP], \qquad (2.3)$$

which results in the following output string:

$$juntar\{C:V, T:IMP, O:3SM\}. \qquad (2.4)$$

Due to the wildcards, rule (3) works for any polite expression in the form *podrías* ("could you") + infinitive verb.

Usually, properties found within brackets are accessed for the object whose name appears immediately to the left of those brackets. However, access to properties for other objects outside the brackets is possible through the use of the dot notation defined as follows. Consider the following string:

```
juntar{C:V, T:IMP, O:3SM} un poco más
```
"put it together" "a bit more"

the collocation *un poco más* ("a bit more") can be transformed into a quantity adverb by the rule

$$un \; poco \; más - - > x\{C:ADV, Q:L\}, \qquad (2.5)$$

which then can be transformed into the verb's property by the rule:

$$A[C:V] \; B[C:ADV, Q] - - > A[Q:B.Q], \qquad (2.6)$$

which means, "if a verb A is followed by an adverb B with some quantity, then add to this verb the property quantity with the same value that it has in B." The latter construction is expressed in (2.6) as B.Q standing for the value of Q in B.

If a property is specified for a wildcard that has no value, then—to match the wildcard—the property must be present, regardless of its value.

Note that, because of their ability to replace other properties, wildcards are not reduced for the unification of properties [111].

2.4.2.4 External Procedures

External procedures that include arguments are formed by a procedure name followed by an argument list:

$$\text{procedure\_name (arg}_1, \text{arg}_2, \ldots, \text{arg}_n),$$

where n is a natural number that may be 0, in which case the procedure has no arguments. Unlike functions, procedures do not return any value. They are executed by the SOC system's engine after the successful application of rules over an utterance. For example,

$$\texttt{move(A,B)}$$

is an external procedure that places object A in position B.

2.4.2.5 Symbolic References to Objects

A scene is an object composed of other objects. In turn, these other objects can be composed of still other objects. For example, catalogs are objects that are composed of elements that are also objects.

Such compositionality permits us to establish nested contexts in order to resolve references to objects based on the scene on which the user's attention is focused at any given moment. Each of the objects inside a scene has properties that can be accessed by our conversion rules by means of tags.

In contrast to grammatical properties, which are described exclusively within our conversion rules, object properties belong to the SOC system and can vary. These properties can be, for example, position, size, components, color, material, density, alterability, shape, and any set of actions that can be applied to the given object.

Labels that begin with `obj_` denote symbolic references to objects. For example, `obj_box231` refers to a particular box that appears in a particular scene.

2.4.2.6 Embedded Functions for Context and Object Reference Handling

Embedded functions, which are a means to provide object reference handling, are discussed in the next section.

2.5   Object Reference and Context Management

For each noun, pronoun, or noun phrase, we need to find a unique symbolic reference to the particular object meant by the user. However, the same expression (as strings of letters) can be used to refer to different particular objects, depending on the context. To transform an expression into a symbolic reference, we should first determine the context for it [158].

To provide context handling, we consider context to be an object (called a "scene object") that contains other objects. Similarly to SQUIRREL [9], in our model, context and object reference are managed by stacks. However, we use three stacks instead of one: subject-object stack (sos), object-object stack (oos), and context (or scene) stack (ss).

A context change occurs when the user shifts his or her attention from the object itself to its components, or vice versa. For example, the user can consider a catalog or objects from this catalog or parts of specific objects from this catalog. Here, we see that catalog objects belong to one context (the catalog), while objects in it belong to another context. Each of these contexts is called a scene.

2.5.1   Embedded Functions for Context and Object Reference Management

Besides standard operations over stacks (push and pop), we can search for objects by property in a given stack (sos, oos, or ss). Embedded functions for objects and context management are listed below. These functions are executed in-line, that is, they are evaluated immediately after application of the rule that generated them in the string and before applying another rule.

Syntactically, embedded functions are denoted by the function name followed by the argument list, which may be empty:

```
object function_name(arg₁, arg₂, ..., argₙ),
```

where n is a natural number (possibly zero). A function must return an object. Table 2.2 shows the embedded functions and procedures used in our formalism.

Using the three stacks, we can define the procedure for searching for the object referenced by the user as follows:

```
            search object in sos
               if it is not found: search object in oos,
                  P1: if it is not found: goto SearchSS
               SearchSS:
      search object in ss,
                  if it is not found: until it is found do:
                         ss.pop();
               repeat SearchSS.
```

Table 2.2 Embedded functions and procedures

Function	Description
push (s, x)	Pushes the object x onto the stack s
object pop (s)	Pops and returns the top object from the stack s
object last (s)	Returns the object from the top of the stack s without popping it
object last (s, p = v)	Searches for the first object with the value v of the property p, starting from the top of the stack s. If no object is found, it returns NIL

2.5.2 Conditional Markers

A conditional marker is a function used for making decisions during rule processing. Its format is

if <condition> then <object$_1$> else <object$_2$> end.

This in-line function returns object$_1$ if the condition is met; it returns object$_2$, otherwise. For example, the procedure SearchSS (above) can be implemented as follows:

```
A[C:ARTDET] B[C:SUST] ->
if sos.last (name = A.name) then sos.last (name = A.name) else
if oos.last (name = A.name) then oos.last (name = A.name) else
    if ss.last (name = A.name) then ss.last (name = A.name)
      else ss.pop () A B end
```

As we can see from this rule, recursion is expressed by rewriting the left-hand side of the rule as the right-hand side, which—in this rule—is expressed as A B in the last line.

2.6 Processing of Sample Queries

Here, we present a rule set that is able to process several utterances in the Spanish language. These utterances (abridged "utt") are inspired by dialogues presented in [157].

utt.1: *¿Me puedes mostrar el catálogo?* "Can you show me the catalog?"

utt.2: *¿Me puedes mostrar el catálogo de objetos formados por esferas?* "Can you show me the catalog of objects made of spheres?"

utt.3:   *A ver, ¿cuál es la diferencia entre el tercero y el cuarto?* "Let me see, what is the difference between the third and the fourth one?"

utt.4:   *¿Me puedes poner el tercero junto al toroide?* "Can you put the third one next to the torus for me?"

2.6.1 Rule Set

Here is the set of rules used to analyze the fragment of utterances presented above. Rule 1 synthesizes procedure P1 for referenced object searching.

```
 1   A[C:ARTDET] B[C:SUST] ->
     if(sos.last(name = A.name),sos.last(name = A.name),
     if(oos.last(name = A.name),oos.last(name = A.name),
     if( ss.last(name = A.name), ss.last(name = A.name),
     ss.pop();A B))))
 2   [C:ARTDET] cuarto{C:ADJ} -> 4{C:SUST}
 3   [C:ARTDET] tercero{C:ADJ} -> 3{C:SUST}
 4   B diferencia entre D y F -> diferencia D F
 5   B[C:SUST,Q] -> if (ss.last(name = B.name),ss.last(name = B.name),
         ss.last(prop = B.Q))
 6   el -> el{C:ARTDET,S:SM}
 7   nextto A[C:OBJ] -> getpos(A); oos.push(A)
 8   cuarto -> cuarto{C:ADJ,S:SM}
 9   diferencia A[C:OBJ] B[C:OBJ] -> diff(A,B);sos.push(A);sos.push(B);
10   el catálogo de objetos formados por esferas -> cat021
11   junto a -> nextto
12   me poder{C:V,T:PRES,S:2S} A[C:V,T:INF] -> A[T:IMP,S:2S,O:1S]
13   mostrar -> mostrar{C:V, T:INF}
14   mostrar{C:V,T:IMP,S:2s} A[C:OBJ] -> show(A); ss.push(A);
15   poner{C:V,T:IMP,S:1S} B[C:OBJ] D[C:POS] -> F=move(B,D);sos.push(F);
16   puedes -> poder{C:V,T:PRES,S:2S}
17   tercero -> tercero{C:ADJ,S:SM,Q:3}
18   catálogo -> catálogo{C:SUST,S:SM,Q:4}
19   poner -> poner{C:V,T:INF,S:1S}
20   al -> a el
```

2.6.2   *Rules in Action*

Now, we can process the utterances presented at the beginning of this section. The number at the left indicates the number of the rule used between one step and the next.

utt1: ¿Me puedes mostrar el catálogo? "Can you show me the catalog?"

```
16. me poder{C:V,T:PRES,S:2S} mostrar el catálogo
13. me poder{C:V,T:PRES,S:2S} mostrar{C:V,T:INF} el catálogo
12. mostrar{C:V,T:IMP,S:2S,O:1S} el catálogo
18. mostrar{C:V,T:IMP,S:2S,O:1S} el catálogo{c:SUST,S:SM}
 6. mostrar{C:V,T:IMP,S:2S,O:1S} el{C:ARTDET,S:SM} catálogo{c:SUST,S:SM}
 1. mostrar{C:V,T:IMP,S:2S,O:1S} objcat01
14. show(objcat01);ss.push(objcat01);
```

utt2: ¿Me puedes mostrar el catálogo de objetos formados por esferas? "Can you show me the catalog of objects made of spheres?"

```
16. me poder{C:V,T:PRES,S:2S} mostrar el catálogo de objetos formados por esferas
13. me poder{C:V,T:PRES,S:2S} mostrar{C:V,T:INF} el catálogo de objetos formados por
    esferas
12. mostrar{C:V,T:IMP,S:2S,O:1S} el catálogo de objetos formados por esferas
10. mostrar{C:V,T:IMP,S:2S,O:1S} cat021
14. show(cat021);ss.push(cat021);
```

utt3: A ver, ¿Cuál es la diferencia entre el tercero y el cuarto? "Let me see, what is the difference between the third and the fourth one?"

```
 4. diferencia el tercero el cuarto
 6. diferencia el{C:ARTDET,S:SM} tercero el cuarto
 6. diferencia el{C:ARTDET,S:SM} tercero el{C:ARTDET,S:SM} cuarto
 2. diferencia el{C:ARTDET,S:SM} tercero 4{C:SUST}
 3. diferencia 3{C:SUST} 4{C:SUST}
 5. diferencia objcat021_03 4{C:SUST}
 5. diferencia objcat021_03 objcat021_04
 9. diff(objcat021_03,objcat021_04); sos.push(objcat021_03); sos.push(objcat021_04);
```

utt4: ¿Me puedes poner el tercer junto al toroide? "Can you put the third one next to the torus for me?"

```
16. me poder{C:V,T:PRES,S:2S} poner el tercero junto al toroide

19. me poder{C:V,T:PRES,S:2S} poner{C:V,T:INF,S:1S} el tercero junto al toroide

 6. me poder{C:V,T:PRES,S:2S} poner{C:V,T:INF,S:1S} el{C:ARTDET,S:SM} tercero junto al
    toroide

20. me poder{C:V,T:PRES,S:2S} poner{C:V,T:INF,S:1S} el{C:ARTDET,S:SM} tercero junto a
    el toroide

 6. me poder{C:V,T:PRES,S:2S} poner{C:V,T:INF,S:1S} el{C:ARTDET,S:SM} tercero junto a
    el{C:ARTDET,S:SM} toroide

11. me poder{C:V,T:PRES,S:2S} poner{C:V,T:INF,S:1S} el{C:ARTDET,S:SM} tercero nextto
    el{C:ARTDET,S:SM} toroide

17. me poder{C:V,T:PRES,S:2S} poner{C:V,T:INF,S:1S} el{C:ARTDET,S:SM}
    tercero{C:ADJ,S:SM} nextto el{C:ARTDET,S:SM} toroide

12. poner{C:V,T:IMP,S2s,O1S} el{C:ARTDET,S:SM} tercero{C:ADJ,S:SM} nextto
    el{C:ARTDET,S:SM} toroide

 3. poner{C:V,T:IMP,S2s,O1S} 3{C:SUST} nextto el{C:ARTDET,S:SM} toroide

 5. poner{C:V,T:IMP,S2s,O1S} objcat021_03 nextto el{C:ARTDET,S:SM} toroide

 1. poner{C:V,T:IMP,S2s,O1S} objcat021_03 nextto obj002; oos.push(obj002);

 7. poner{C:V,T:IMP,S2s,O1S} objcat021_03 getpos(obj002); oos.push(obj002);

15. move(objcat021_03,getpos(obj002));sos.push(obj003); oos.push(obj002);
```

In the last line, the object is copied when it is moved from one context to another.

2.7   Conclusions

In this chapter, we have presented a system that can derive one or more specific computer instructions from a request by the end user. Objects referenced within this request are translated to symbols by a rewriting rules grammar with property modification, wildcard substitution, in-line functions, and the use of special context objects called "scenes." Instructions are then executed by an external system.

The system can be used for computer tasks that meet the following conditions: the user and the computer share common contexts, which are visualized, and the application domain is limited. We have presented examples for the task of placing objects in a three-dimensional canvas. This work can be extended to cover not only abstract geometric objects, but also world objects within a modeled world. In addition, by using special functions, this system can be enhanced to allow new rules to be created by previous rules, thus automating dynamic grammar extension through dialogue.

SOC systems have characteristics that allow them to directly translate natural language sentences into computer instructions. In an SOC system, the language used is imperative; objects are previously defined and can be combined to create new ones; objects have properties; objects are always present; a spatial, common representation exists visually; and a limited number of actions exist over these objects.

Given these characteristics, we have shown how such translations can be done with the DTG. We have presented both a framework based on that grammar and a mechanism for object reference and context management. The problem of resolving object references is solved within this framework through a context stacks mechanism and conditionals embedded in the DTG's rules.

This system can be extended to allow for new rules to be created out of existing ones. In this way, it can be dynamically extended through dialogue. The development of this idea is a topic for future work.

Chapter 3
Second Approach: Constituent Grammars

In this chapter, we tackle sentence analysis using the constituent approach. This approach has two problems. The first is the difficulty of extracting information about characters and actions from factual reports such as news articles, web pages, and circumscribed stories. To construct the structure of a sentence, there should be interaction with previously acquired knowledge. In turn, such knowledge should be expressed in a structured way so that simple inferences may be used when necessary. We will deal with this problem in detail in Sect. 3.1.

The second problem is the difficulty of obtaining semantic indices from several sentences. Currently, widely known formalisms such as HPSG [175] do not consider this problem by themselves. We will discuss that in Sect. 3.2.

3.1 Representation Using Typed Feature Structures

We propose extracting information about characters and actions from a self-contained story, such as news reports. Such information is stored in structures called "situations." We will show how these situations can be constructed by unifying the constituents of sentence analysis with knowledge previously stored in typed feature structures. These situations can in turn be used in the form of knowledge. This combination of situations constructs a supra-structure that represents the understanding of a factual report and that can be used to answer questions about facts and their participants.

© Springer International Publishing AG 2018
A. Gelbukh and H. Calvo, *Automatic Syntactic Analysis Based
on Selectional Preferences*, Studies in Computational Intelligence 765,
https://doi.org/10.1007/978-3-319-74054-6_3

3.1.1   Introduction

Factual reports are texts in which facts are described in an ordered manner; each participant of these facts is circumscribed within the text. These characteristics allow us to apply techniques for dealing knowledge of practical complexity. By "practical complexity," we mean mid-term feasible applications. For example, understanding a story can be undertaken in several ways, and one element that must be taken into account is the amount of previously acquired knowledge necessary for understanding the story. Minsky [137] includes an example of this type of understanding:

> There was once a Wolf who saw a Lamb drinking at a river and wanted an excuse to eat it. For that purpose, even though he himself was upstream, he accused the Lamb of stirring up the water and keeping him from drinking.

Minsky argues that the key to understanding this text is to realize:

1. The lamb's stirring the water produces mud,
2. If water has mud, it cannot be drunk,
3. If the wolf is upstream, the fact that the lamb stirs the water does not affect the wolf, and so
4. The wolf is lying.

However, these inferences require quite a large structured knowledge system for a machine, and the construction of such a general knowledge system is not our goal at this point. First, we need to solve tasks at a lower level:

1. Identification of characters, places, and objects in the story.
2. Identification of the described actions.
3. Identification of actions that are not done but are mentioned within the story.
4. Formulation of the arguments for each action. These arguments can be seen as answers to "wh" questions: who, where, what, when, why, and whom. Each action with its arguments is a structure that we call "situation."
5. Establishment of the temporal sequence of situations corresponding to the story's flow.

Following this approach, for the passage of the wolf and the lamb, we find that:

1. The characters are the Wolf and the Lamb.
2. The places are the River and Upstream.
3. The situations are:

   - The Wolf sees the Lamb.
   - The Lamb drinks.
   - The Lamb is *in* the River.
   - The Wolf wants (to eat the Lamb).
   - The Wolf is *in* Upstream.
   - The Wolf accuses the Lamb *of* (stirring *the* Water).
   - The lamb *does* not let (the Wolf drink the Water).

Each string of words in parenthesis is also a situation, but note that those situations do not necessarily occur. In this case, the Wolf's eating the Lamb does not occur (nor do we know that it will occur). Similarly, the other two parenthetical situations—the Lamb's stirring the Water and the Lamb's not letting the Wolf drink the Water—do not actually occur in this story.

(Capitalized words point to particular instances of characters, places, and objects in this particular story.)

3.1.2 Representing Situations with Typed Feature Structures

In order to construct and represent situations, we propose use of typed feature structures (TFSs). This formalism permits us to cover every level of linguistic description [43]: basic sentence type (PoS) construction, intermediate sentence type construction (e.g., structures for specifying individuals), situations with complements construction, and, finally, story structure construction.

We can represent a situation as a feature structure, as shown in Fig. 3.1. This representation is an attribute value matrix (AVM). We represent attributes in uppercase letters and values in lowercase ones.

For instance, *sit_thing* indicates that the values for WHAT and WHY can be a *situation* and *thing*. TIME has a numeric value that corresponds to the sequence in which situations are mentioned. OCCURS is the feature that indicates whether a situation occurs within the story.

Fig. 3.1 AVM (attribute value matrix) for the type *situation*

$$
\begin{bmatrix}
situation & \\
\quad \text{ACT} & action \\
\quad \text{TIME} & n \\
\quad \text{WHO} & individual \\
\quad \text{WHAT} & sit\_thing \\
\quad \text{WHERE} & place \\
\quad \text{WHOM} & individual \\
\quad \text{WHY} & sit\_thing \\
\quad \text{WITH} & thing \\
\quad \text{NEG} & *boolean* \\
\text{OCCURS} & occ
\end{bmatrix}
$$

The fact that feature structures are typed permits us to handle object hierarchies, such as stating that a man is a human, that humans are individuals, and that, therefore, humans can be participants in an action as values for WHO and/or WHOM.

3.1.2.1 Interaction Between Syntax and Knowledge

Before we explain how situations are constructed, we will discuss briefly the interaction between syntax and knowledge.

Traditionally, TFSs have been used mostly for syntax analysis, whereas frame-based systems are mostly used to handle knowledge. Examples of the use of these formalisms are HPSG [175], a well-known formalism that combines the use of generative grammars with the benefits of TFS, and NEOCLASSIC [153], a frame-based knowledge representation system.

We believe that in order to successfully construct a situation, these two traditionally separate stages need to be blended so that interaction between them is possible. The formalism we choose to represent both syntax and knowledge is TFS, because it shares characteristics with frame-based systems:

1. Both frames and TFS are organized in hierarchies.
2. Frames are composed out of slots (equivalent to feature structures' attributes) for which fillers (equivalent to feature structures' values or references to other frame-feature structures) must be specified or computed [144].
3. Both frames and TFS are declarative.
4. TFS's logic is similar to the logic used by frames: description logics.

Description logics are part of first-order logics and are used by frame-based systems for reasoning. In description logics, it is possible to construct a hierarchy of concepts from atomic concepts and attributes, usually called "roles" [145]. The only difference between the feature logics used by TFS and the description logics used by frame-based systems is that feature logics' attributes are single-valued, while description logics' attributes are multi-valued. This may seem like a slight difference, but it could be the difference between decidable and undecidable reasoning problems [146].

3.1.2.2 Construction of Situations

The linguistic knowledge building (LKB) system is a programming environment for TFS. LKB follows the formalism introduced by Shiber [180].

Although LKB has been most extensively tested with grammars based on HPSG [175] (e.g., ERG from the LinGO project [60]), LKB is intended to be framework-independent [59].

LKB is used mostly for the lexical parsing of sentences; however, it can also be used for storing and interacting with knowledge.

Currently, to deal with a semantic representation, LKB makes use of a grammar that has special markers (e.g., LISZT and HANDEL) to build a minimal recursion semantics (MRS) [63] representation. MRS produces a flat logical representation intended mainly for handling transfer and quantifier phenomena. MRS output is mainly suited for translation and has been successfully used in the Verbmobil project [42]. However, for our purposes, MRS drops specific syntactic information that would allow us to identify the grammatical role that each constituent plays, thus making it difficult to determine if a constituent fills the WHO or the WHOM slot, for example.

As we proposed in Sect. 3.1, we will use TFS to construct a situation. The construction of situations corresponds to each sentence.

To handle *situations* as TFS in LKB, we establish the types shown in Fig. 3.2, along with their corresponding hierarchy, using standard LKB notation for lists, TFS, and unification (i.e., &). We assume that types enclosed in asterisks are predefined, with *top* being the most general type available in the type hierarchy.

To illustrate how situations are constructed, we will use the Wolf and Lamb example from [137]. A fragment of that story is reproduced here for your convenience:

> There was once a Wolf who saw a Lamb drinking at a river and wanted an excuse to eat it. For that purpose, even though he himself was upstream, he accused the Lamb of stirring up the water and keeping him from drinking.

The example we present illustrates the construction of situations. Therefore, syntax and other phenomena—with their inherent complexity—are not covered here. Syntax analysis is addressed in this example as pattern matching. For larger scale systems, formalisms such as HPSG can be used within this approach since such formalisms are capable of handling TFS.

To construct a situation, we assume that the system previously obtained information about the possible roles that each entity can have. For example, "River" and "Upstream" are places, "Wolf" and "Lamb" are individuals, and "Water" is an object (see Fig. 3.3).

Entities can be formed by more than one word. We do not know a priori any of the possible properties these entities may have (e.g., *big* Lamb, a Lamb *named* Dolly, etc.). These properties will be filled in as the story is analyzed.

Knowledge of whether or not a situation occurs is important to understanding the flow of the story. In the example of the Wolf and the Lamb, the situation of the Lamb stirring the Water and thus keeping the Wolf from drinking does not really occur; this is a situation mentioned as a consequence of the Wolf wanting an excuse to do something. Thus, to define whether a situation occurs, we consider that when a situation is subordinated by other situation, the subordinated situation does not occur.

We will analyze the fragment of the story presented above word-by-word following a specific order. Note that feature logic is declarative; thus, this analysis could be done in any order and still yield the same results.

```
n := *top*.

occ := *boolean*.
occ_if := occ &
    [IF situation].

whatsit := situation &
    [WHAT sit_thing].

whysit := situation &
    [WHY  sit_thing].

sit_thing := situation
& thing.

tpi := *top* &
    [ORTH string].

thing := tpi.
place := tpi.
individual := tpi.
action := tpi.

situation := *top* &
    [ACT action,
     TIME n,
     WHO individual,
     WHERE place,
     WITH thing,
     NEG *boolean*,
     WHOM individual,
     OCCURS occ].

story := *top* &
    [INDIVIDUALS
*list*,
     PLACES *list*,
     OBJECTS *list*,
     ARGS *list*].
```

```
#wolf [individual]
#lamb [individual]
#river [place]
#upstream [place]
#water [object]
```

We will begin by analyzing the first sentence (3.1):

There was once a wolf who saw a lamb drinking at a river and wanted an excuse to eat it.

$$(3.1)$$

The first words of (3.2) match a pattern that introduces the `wolf` as an individual: `there + was + once + a + individual`. This pattern then leads to the representation shown in (3.3).

there was once a Wolf (3.2)
[individual

NAME wolf (3.3)

ORTH "wolf"]

This structure can be combined with a corresponding structure in the knowledge base (implemented as TFS) to find the possible properties of wolves in general.

To avoid rewriting the feature structures we identify in this analysis, we instead write a reference to those structures using LKB notation, in which labels begin with #.

#wolf [individual

    NAME wolf (3.4)
      ORTH "wolf"].

We can then write the sentence being analyzed as (3.5):

#wolf who saw a Lamb drinking at a river (3.5)

A feature structure of type *individual* followed by a lexeme *who*, makes *who* absorb the individual. The rule doing this is (3.6):

individual_who := individual &
 [NAME #1,
   ORTH #2,
   ARGS < individual & [NAME #1, (3.6)
         ORTH #2],
      lexeme_who]>].

The sentence now becomes

#wolf saw a Lamb drinking at a river. (3.7)

We then turn our attention to *Lamb drinking at a river*. This is another situation, but first we must add the `lamb` individual to our story.

$$\#lamb \; [individual$$
$$NAME \; lamb \qquad\qquad (3.8)$$
$$ORTH \; "lamb"].$$

a `Lamb drinking at a river` is then converted into

$$\#lamb \; drinking \; at \; a \; river. \qquad\qquad (3.9)$$

The previously defined lexicon (see Fig. 5.3) provides the information that `river` can be a place. However, `river` is not restricted to only one category; in case several choices are available, unification will help select the correct one(s). `river` is then considered as

$$\#river \; [place$$
$$NAME \; river \qquad\qquad (3.10)$$
$$ORTH \; "river"]$$

We do not show here the details of a reference resolution mechanism that could establish the difference between "a river" and "the river" according to previously introduced entities. Instead, for this work, we assume that each time an entity is mentioned, its feature structure equivalent is introduced. When the supra-structure story is formed, two feature structures (FSs) corresponding to the same entity will unify. If two FSs of the same kind have conflicting particular characteristics (such as, red river and blue river), unification will fail and then two different entities will be considered.

#river can be later unified with a knowledge base so that the system is able to infer that #river is made of #water. For simplicity in this example, we assume that this kind of information has not been implemented.

Returning to the analysis of (3.9), we can verify from our lexicon that `drinking` unifies with the type *action* (verb). We will call the feature structure for this particular action #drink (3.11), thus obtaining (3.12).

$$\#drink \; [action$$
$$NAME \; drink$$
$$TENSE \; gerund \qquad\qquad (3.11)$$
$$ORTH \; "drinking"]$$

$$\#lamb \; \#drink \; at \; \#river \qquad\qquad (3.12)$$

Now we can apply the FS rule that creates a situation when the sequence: `individual`, `action`, "at," `place` is found:

$$
\begin{array}{l}
\texttt{[situation} \\
\texttt{ACT \#2} \\
\texttt{WHO \#1} \\
\texttt{WHAT} \\
\texttt{WHERE \#3} \\
\texttt{ARGS <\#1, \#2, lexeme\_at, \#3 >]}
\end{array}
\qquad (3.13)
$$

Exceptions to rule (3.13) can be handled as additional constraint rules. Applying this rule, we have situation #s2:

$$
\begin{array}{l}
\texttt{\#s2 [situation} \\
\texttt{ACT drink} \\
\texttt{who \#lamb} \\
\texttt{WHAT} \\
\texttt{WHERE \#river]}
\end{array}
\qquad (3.14)
$$

We return to the main sentence (3.7), substituting the last situation we have just found:

$$
\texttt{\#wolf saw \#s2.} \qquad (3.15)
$$

This then forms another situation:

$$
\begin{array}{l}
\texttt{\#s1 [situation} \\
\texttt{ACT see} \\
\texttt{WHAT \#s2]}
\end{array}
\qquad (3.16)
$$

Finally, the first sentence is a situation:

$$
\texttt{\#s1} \qquad (3.17)
$$

`#s1` has a subordinated situation, `#s2`. The rest of the story fragment of the Wolf and the Lamb can be analyzed in a similar way. The entities consulted from the lexicon are shown in Fig. 3.2 (LKB types for representing situations and stories) and the story structure obtained after this analysis is shown in Fig. 3.4 (Feature structure for the story fragment of the Wolf and the Lamb).

```
story & [
        INDIVIDUALS   <#wolf & wolf1, #lamb & lamb1>,

        PLACES        <#river & river1, #upstream &
                                        upstream1>,

        OBJECTS       <#water & water1>,

        SITUATIONS    <#S1 [situation        #S2 [situation
                            TIME 1                TIME 1
                            ACT see               ACT drink
                            WHO #wolf             WHO #lamb
                            WHAT #s2              WHAT (liquid)
                            OCC yes],             WHERE #c
                                                  OCC yes],

                       #s3 [situation        #s4 [situation
                            TIME 2                TIME 2
                            ACT want              ACT eat
                            WHO #wolf             WHO #wolf
                            WHAT #s4              WHAT #lamb
                            OCC yes],             OCC no],

                       #s5 [situation        #s6 [situation
                            TIME 3                TIME 4
                            ACT is                ACT accuse
                            WHO #wolf             WHO #wolf
                            WHERE #upstream       WHAT #s8
                            OCC yes],             WHY #s3
                                                  OCC yes],

                       #s7 [situation        #s8 [situation
                            TIME 4                TIME 4
                            ACT accuse            ACT stir
                            WHO #wolf             WHO #lamb
                            WHAT #s9              WHAT #water
                            WHY #s3               OCC no],
                            OCC yes],

                       #s9 [situation        #s10 [situation
                            TIME 4                 TIME 4
                            ACT let                ACT drink
                            WHO #lamb              WHO #wolf
                            WHAT #s10              WHAT (liquid)
                            NEG true               OCC no] >]
                            OCC no],
```

Fig. 3.4 Feature structure for the story fragment of the Wolf and the Lamb

3.1.3 Minsky's Frames and Situations

Minsky argues in [137] that frames are like a network of nodes and relationships:
the top levels of a frame are fixed and represent things that are always true about a
supposed situation, while lower levels have terminals (slots) that must be filled by
specific instances or data. Conditions specified by markers require that a slot
assignment is a person, object, or pointer to a sub-frame of a certain type.

A terminal that has acquired a *feminine person* marker will reject pronominal *masculine* assignments. In this sense, Minsky's frames are very similar to a feature structure. Each frame could be regarded as a feature structure, with slots being the values of the attributes in the attribute-value structure. However, there is an important difference among Minsky's frames and our view of situation representation.

Minsky talks about frames as data structure to represent a stereotyped situation, like being in a special kind of room or going to a child's birthday party. Minsky considers that frames must contain information about how they must be used, about expectations, and about what to do if expectations are not met. In contrast, we consider a situation as a simple transitory unit of the state of things within a story. Consider the sentence: *The man wants to dance with Mary*. This sentence contains two situations: (Situation 1) *wants*. Who? *The Man*. What? refers to situation 2. Occurs? Yes. (Situation 2) *dance*. Who? *The Man* (the same man), (with) Whom? *Mary*? Occurs? No. In these situations, we do not consider (in contrast with Minsky) information about how to use a frame, expectations, or what to do if expectations are not met.

3.2 Adding a Knowledge Base

In this section, we show how a knowledge base can be constructed semi-automatically from a text using typed feature structure (TFS)-based formalisms such as HPSG. This knowledge base can be consulted in subsequent analyses in order to resolve intersentential co-references.

Understanding a text implies being able to identify the entities described in it, the properties of those entities, and the situations in which they participate. Modern grammars usually specify such information using reference indices. For example, the entry for *give* in HPSG is defined as [175].

$$
\begin{bmatrix}
\textit{dtv-lxm} \\
\text{ARG-ST} \quad \langle [\,]_i, [\,]_j, [\,]_k \rangle \\
\text{SEM} \quad
\begin{bmatrix}
\text{INDEX} \quad s \\
\text{RESTR} \quad \left\langle
\begin{bmatrix}
\text{RELN} & \text{give} \\
\text{SIT} & s \\
\text{GIVER} & i \\
\text{GIVEN} & j \\
\text{GIFT} & k
\end{bmatrix}
\right\rangle
\end{bmatrix}
\end{bmatrix}
$$

Here, in the semantic section (**SEM**) of the definition, the entities participating in situation s are referred to by the indices i, j, and k. Different entities are referred to by different indices, and the same entity by the same index (co-reference).

However, the implementations we are aware of maintain such correspondence only within one sentence. For each new sentence, the count of indices restarts from 1, thus destroying the one-to-one correspondence between entities and indices: the entity referred to by index 1 in one sentence has nothing to do with the entity referred to by index 1 in the next sentence. Thus, to maintain semantic coherence within the discourse, it is important to correlate all indices that refer to the same entity throughout the text.

To do so, we propose a mechanism that creates and maintains semantic structures separately from sentence analysis. We store these structures in a knowledge base that is built alongside text parsing. In this base, the structures are in a one-to-one correspondence with the entities mentioned throughout the text.

Apart from representing the semantics of the text, this knowledge base can be consulted during text analysis to resolve co-references: when an entity with certain properties is mentioned in the text in the context implying co-reference (e.g., when the entity appears with a definite article), we look in the knowledge base for a suitable entity with the same or compatible properties that was defined earlier. Various heuristics are applied and different sources of evidence are considered in making the final decision regarding the presence or absence of a co-reference.

Discussion of the co-reference resolution heuristics is out of the scope of this section: here, we only discuss the form of representation used for entities and situations in our knowledge base. In Sect. 3.2.1, we consider the TFS formalism for parsing and knowledge representation. In Sect. 3.2.2, we explain the desired contents of our knowledge base. In Sect. 3.2.3, we discuss how it is built.

3.2.1 TFS as Knowledge Representation

Since we want to combine the functionality of HPSG-like grammars with a knowledge base, it is desirable to use a single formalism for both sentence analysis and knowledge representation.

For knowledge representation, systems based on description logics (DL) are traditionally used. These systems are also known as "terminological logics systems" or "KL-ONE-like systems," e.g., NEOCLASSIC [153], BACK [103], CRACK (online) [26], FaCT [12], LOOM [121], and RACER [98]. However, such formalisms are not designed for sentence analysis.

On the other hand, the same formalism—TFS—that is used in HPSG-like grammars for sentence analysis can be used for knowledge representation—specifically, for building the knowledge base while parsing text and even for implementing reasoning [43]. The logic implemented in TFS by means of unification is called feature logic (FL). As in DL, in FL it is possible to construct a concept hierarchy from atomic concepts and attributes usually referred to as *roles* in DL. The main difference between FL and DL is that FL's attributes are single-valued, while attributes in DL are multi-valued [145]; however, this is a minor difference since, in FL, attributes can be lists of values.

For example, in the DL system NEOCLASSIC, one can create an individual using the function createIndividual (sandy person). This can be represented by the TFS sandy, which is a subtype of person. Afterward, addToldInformation (sandy fills (has dress)) can be seen as an operation on the TFS sandy, thus adding a feature. The resulting TFS is

sandy
[HAS dress].

For retrieval, getInstances (<concept>) is used, where <concept> (e.g., HASDRESS) can be defined as createConcept (HASDRESS fills (has dress)). This way, the command getInstances (HASDRESS) is equivalent to unifying all available instances with the TFS [HAS dress], thus obtaining the whole TFS sandy.

3.2.2 Structure of the TFS Knowledge Base

Now that we have shown that TFS can be used for knowledge representation, we will discuss the structure of the knowledge base we want to obtain from a text.

An HPSG-like grammar has a lexicon that relates a word string with the types (parts of speech) to which the string can be mapped. The gradual combination of terms constructs parts of speech, which in turn form syntagmata.

At an appropriate point, we convert these terms into entities—TFS representations of animate or inanimate, real or abstract objects. These entities are added to the knowledge base.

Figure 3.5 shows the structures that result in the knowledge base after the text fragment is analyzed.

$$\text{There is a big, red bookshelf}_i \text{ in the living room}_j. \text{ John}_{i'}\text{'s books}_{j'} \text{ are neatly placed on it}_k \qquad (3.18)$$

References to entities are marked here with the corresponding sentence analysis indices, which make clear the correspondence between TFS structures and the text. However, these indices do not point to letter strings; rather, they are sequential numbers produced by sentence analysis. In Fig. 3.5, strings such as *John*, *bookshelf*, and *it* are included only for clarity and are not a proper part of the structures.

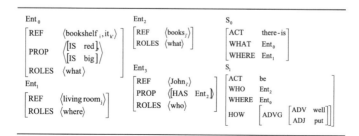

Fig. 3.5 TFS structures extracted from text (3.1)

The name of a TFS is formed by its type and indexed by a number. Ent_0, …, Ent_3 are entities; note that they store the role they are given when used in situations. In Fig. 3.5, ACT stands for action, REF for reference, ADVG for adverbial group, ADV for adverb, and ADJ for adjective.

S_0 and S_1 are situations. Situations are formed by the attributes **ACT** (action), **WHO, WHAT, WHERE, WHOM, WHY,** and **WITH,** among others.

Note that sometimes the semantic relationship cannot (easily) be obtained from the immediate context. For example, *John's books* does not necessarily refer to the books that John *has*, but perhaps to those that he has *written*. A possible way to resolve such ambiguities is to ask a human operator; other possible ways are not discussed here.

Semantic representation of text, as in Fig. 3.5, allows us to search for entities mentioned in that text by a given set of properties. This can be used both for co-reference resolution during text analysis and for answering questions. For example, a dialogue system allows users to ask: *Is there a bookshelf in the living room? Where are John's books?* The answers to these questions can be easily found by unifying the properties mentioned in the questions with the objects stored in the knowledge base.

3.2.3   Building a TFS Knowledge Base

We have so far seen how four entities (*a bookshelf*, *the living room*, *the books*, and *John*) extracted from two sentences can be represented. Now, we will briefly describe the mechanism that allows a TFS parse system to construct and use such representations.

To maintain the knowledge base (KB), three functions are used: INTRO, ADD, and GET. The objects introduced into the base with the function INTRO persist in a scope that is broader than a single sentence and can be modified with ADD or retrieved with GET while analyzing other sentences. Since the entities are represented as TFS, unification is the only operation that we use in these functions.

3.2.3.1   Function **INTRO**

This function adds an entity to the KB and returns a pointer to the newly added entity. This pointer is used as a term in the TFS rules. The argument for INTRO is a <TFS_description>. A <TFS description> is a TFS with the specification of the attribute's values in the notation **ATTRIBUTE:value.** This specification can be incomplete (under-specified). If some value is another TFS, we enclose it in parentheses.

INTRO is similar to NEOCLASSIC's **createConcept.** An example of using this function is:

INTRO(IND:(REF:i))

In this example, i will be taken as the index for the current individual's feature REF.

3.2.3.2 Function ADD

With this function, we can add attributes to entities previously created in the KB. ADD's argument is a <TFS_description>, where the TFS of the TFS description is an instantiated TFS previously created in the KB with INTRO. For this function, the <TFS_description> must be complete (all values at every level must be specified). If the entity referenced by the <TFS_description> has already filled the attribute that we are attempting to add but its value is different, the attribute's value is converted into a list that contains both values. If the attribute's value is already a list and does not contain the value we are adding, then the new value is appended.

List elements can be later selected by standard unification methods without considering the order of these elements within the list. The following example demonstrates the addition of an adjective as a property of an individual:

ADD(IND:(PROP:(ADJ)))

3.2.3.3 Function GET

This function returns the entity or entities that unify with the <TFS_description> provided. For example,

GET(S:(WHO:IND1,WHOM:IND2))

obtains all situations where the agent unifies with IND1 and the beneficiary unifies with IND2. IND1 and IND2 are terms that correspond to specific entities derived from a previous analysis.

3.3 Conclusions

We have shown that, for sentence analysis, the typed feature structure (TFS) formalism permits us to address different levels to better understand a story. TFSs are a well-studied formalism that guarantees the computability of its logic. The groundwork we have presented allows situations to be extracted from a factual report so that it is possible to ask simple questions about the text—such as who did something or where she or he did it. This can be used in a web-query system to obtain relevant results about events described in a factual report.

We have also proposed using a knowledge base with persistent objects (entities and situations) in order to maintain co-references across sentences in HPSG-like grammatical formalisms. This knowledge base is built from the text in a (semi)-

automatic way. Entities are available during the whole analysis of a text (rather than for only one sentence) and can also be used after the text has been analyzed, e.g., to answer questions or as a semantic representation of the text.

Adhering to Minsky's frames approach allows us to analyze individuals throughout a story so that characters' behavior can be generalized in a model in order to predict their reactions and interactions, thus tending toward common sense acquisition and expectations in the sense of Minsky's frames. However, for practical analysis of great quantities of text, a large grammar and a large collection of TFSs become unwieldy; thus, we need to compare other approaches for efficient sentence analysis.

Chapter 4
Third Approach: Dependency Trees

After exploring several approaches and representational structures in the previous two chapters, we found that the formalism that best suits our needs is the dependency tree representation. Thus, in this chapter, we present a parser that is based on a dependency tree. This parser's algorithm uses heuristic rules to infer dependency relationships between words, and it uses word co-occurrence statistics (which are learned in an unsupervised manner) to resolve ambiguities such as PP attachments. If a complete parse cannot be produced, a partial structure is built with some (if not all) dependency relations identified. Evaluation shows that in spite of its simplicity, this parser's accuracy is superior to existing available parsers (as tested for use with Spanish). Though certain grammatical rules, as well as lexical resources, are specific for Spanish, the suggested approach is language-independent.

4.1 Introduction

As we have seen, the two main approaches to syntactic analysis are those oriented to the constituency and dependency structures, respectively. In the constituency approach, the structure of the sentence is described by grouping words together and specifying the type of each group, usually according to its main word [50]:

$$\Big[[\textit{The old man}]_{\text{NP}} \big[\textit{loves} [\textit{a young woman}]_{\text{NP}} \big]_{\text{VP}} \Big]_{\text{S}}$$

Here, NP stands for noun phrase, VP for verb phrase, and S for the whole sentence. Such a tree can also be graphically represented:

© Springer International Publishing AG 2018

A. Gelbukh and H. Calvo, *Automatic Syntactic Analysis Based on Selectional Preferences*, Studies in Computational Intelligence 765, https://doi.org/10.1007/978-3-319-74054-6_4

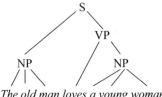

The old man loves a young woman

In this case, the nodes stand for text spans (constituents) and the arcs for "consists of" relationships.

In the dependency approach, words are considered "dependent" from, or modifying of, other words [133]. A word modifies another word (and is called the "governor") if it adds details to that word, and the whole combination inherits the syntactic (and semantic) properties of the governor: *old man* is a kind of *man* (not a kind of *old*); *man loves woman* is a kind of (situation of) *love* (not, say, a kind of *woman*). Such dependency is represented by an arrow from the governor to the governed word:

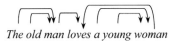

The old man loves a young woman

In graphical form:

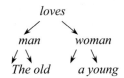

where the arcs represent the dependency relationships between individual words, the words of the lower levels contributing details to those of the upper levels while preserving the syntactic properties of those words.

In spite of a 40-year discussion in the literature, there is no consensus as to which formalism is better. Though combined formalisms such as HPSG [175] have been proposed, these seem to bear both the advantages and disadvantages of the heritage approaches they represent, thus impeding their widespread use in natural language processing. Most likely, the usefulness of one of the two approaches depends on the specific task at hand.

Our motivation was two-fold for this work. Our first was the study of the lexical compatibility of specific words—in particular, the compilation and use of a dictionary of collocations (stable or frequent word combinations) such as *eat bread* or *deep sleep,* as opposed to **eat sleep* and **deep bread* [18]. Such combinations were shown to be useful for tasks ranging from syntactic analysis [209] and machine translation [22] to semantic error correction [23] and steganography [15]. The dependency approach to syntax seems much more appropriate than the constituency approach for such a task.

Our second motivation was the construction of a semantic representation of text, even if partial, for a range of applications from information retrieval and text mining [140, 141] to software specifications [68]. All known semantic approaches—such as conceptual graphs [183], minimal recursion semantics [61], and semantic networks [133]—roughly resemble a set of predicates, where individual words represent predicates or their arguments (which, in turn, can be predicates). The resulting structures are in much closer direct correspondence with a dependency tree than they are with a constituency tree of the same sentence; thus, the dependency syntax seems more appropriate for direct translation into semantic structures. Specifically, the dependency structure makes it much easier to match—say in information retrieval—paraphrases of the same meaning (such as active/passive voice transformation) or to transform one such synonymous structure into another.

In addition, we found that a dependency parser can be much more easily made robust than can a constituency parser. Known approaches to dependency parsing cope much easier with both incomplete grammars and ungrammatical sentences than do standard approaches to context-free parsing.

Indeed, a standard context-free parser builds structures incrementally so that failure to construct a constituent means it is impossible to construct all further constituents that should have contained that one. What is more, an incorrect decision in an early stage of parsing leads to completely or largely incorrect final results.

In contrast, with dependency parsing, the selection of a governor for a given word or the decision of whether the two given words are connected by a dependency relation is much more (though not at all completely) decoupled from corresponding decisions on other pairs of words. This means it is possible to continue parsing even if some decisions cannot be made successfully. The resulting structure may prove to be incomplete (with some relationships missing) or not completely correct (with some relationships wrongly identified). However, an incorrect decision on a particular pair of words usually does not cause a snowball effect of cascaded errors in later steps of the parsing process.

In this chapter, we present DILUCT, a simple robust dependency parser for Spanish. Though some specific rules, lexical resources, and preprocessing tools are specific for Spanish, the general framework is language-independent. An online demonstration and the system's source code are available online.[1]

The parser uses an ordered set of simple heuristic rules to iteratively determine the dependency relationships between words not yet assigned to a governor. In case of certain types of ambiguities, word co-occurrence statistics [which are gathered in an unsupervised manner either from a large corpus or from the Web (through querying a search engine)] are used to select the most probable variant. No manually prepared treebank is used for training.

We evaluated the parser by counting the number of correctly identified dependency relationships on a relatively small treebank. Experiments showed that the

---

[1] www.diluct.com

accuracy of our system is superior to that of existing Spanish parsers, such as TACAT (48) and Connexor.

The rest of this section is organized as follows. In Sect. 4.2, we discuss existing approaches to dependency parsing that have influenced our work. In Sect. 4.3, we present our algorithm; in Sect. 4.3.3, we provide evaluation results.

4.2 Related Work

The dependency approach to syntax was first introduced by Tesnière [114] and then further developed by Mel'čuk [133], who used it extensively in his meaning ⇔ text theory [133, 185] in connection to semantic representation as well as to a number of lexical properties of words, including lexical functions [21, 131].

One of the first serious attempts to construct a dependency parser that we are aware of is the syntactic module of the English-Russian machine translation system ETAP [4]. That parsing algorithm consists of two main steps:

1. Every individual word pair that could have a plausible dependency relationship is identified.
2. So-called filters remove links that are incompatible with other identified links.
3. Of the remaining potential links, a subset forming a tree (namely, a projective tree except for certain specific situations) is chosen.

In ETAP, the grammar (a compendium of situations where a dependency relationship is potentially plausible) is described in a specially developed specification language that describes the patterns for which to search in the sentence and the tree-constructing actions that are to be performed when such a pattern is found. Both the patterns and the actions are expressed in a semi-procedural way by the grammar interpreter's use of numerous built-in functions (some of which are language-dependent). An average pattern-action rule comprises 10–20 lines of tight code. To the best of our knowledge, no statistical information is currently used in the ETAP parser.

Our work is inspired by this approach; however, our main design is different from that of ETAP in several ways. First, our parser is meant to be much simpler, even if at the cost of inevitable loss of accuracy. Second, we do not rely on complex and detailed lexical recourses. Third, we do rely on word co-occurrence statistics, which we believe compensates for the lack of a complete grammar.

Indeed, Yuret [209] has shown that co-occurrence statistics (more precisely, a similar measure that he calls *lexical attraction*) alone can provide enough information for highly accurate dependency parsing, with no handmade grammar at all. His algorithm selects the projective tree that provides the highest total value of lexical attraction of all connected word pairs. However, his approach relies on large quantities of training data (though the training, itself, is unsupervised). In addition, his approach can only construct projective trees (a tree is called "projective" if it has no crossing arcs in the graphical representation shown in Sect. 4.1).

We believe that a combined approach using both a simple handmade grammar and word co-occurrence statistics (that are learned in an unsupervised manner from a smaller corpus) provides a reasonable compromise between accuracy and practical feasibility.

On the other hand, the mainstream of current research on dependency parsing is oriented to formal grammars [65]. In fact, the HPSG grammar [159] was perhaps one of the first successful attempts to, in effect, achieve a dependency structure (which is necessary both for using lexical information in the parser itself and for constructing semantic representations) by using a combination of constituency and dependency machinery. As we have mentioned, low robustness is a disadvantage of non-heuristically-based approaches.

Three syntactic parsers with realistic coverage that are available for the Spanish language are the commercially available XEROX parser[2] and Connexor Machinese Syntax[3] as well as the freely available parser TACAT.[4] We used the latter two systems in order to compare their accuracy with that of our system. Only Connexor's system is really dependency-based, relying on the functional dependency grammar formalism [188]; the other two are constituency-based.

4.3 Algorithm

Following the standard approach, we first pre-process the input text, which basically includes tokenizing, sentence splitting, tagging, and lemmatizing, and then, we apply the parsing algorithm proper.

4.3.1 Preprocessing

4.3.1.1 Tokenization and Sentence Splitting

The text is tokenized into words and punctuation marks and then split into sentences.

We currently do not distinguish between punctuation marks; thus, each mark is substituted with a comma (in the future, we will consider different treatments for different punctuation marks).

Two compounds of article and preposition are split: $del \rightarrow de\ el$ (of the) and $al \rightarrow a\ el$ (to the).

---

[2]Which used to be on www.xrce.xerox.com/research/mltt/demos/spanish.html, but seems to have been recently removed.

[3]www.connexor.com/demo/syntax

[4]www.lsi.upc.es/~nlp/freeling/demo.php

Compound prepositions represented in writing as several words are joined into one word. For example: *con la intención de* (in order to), *a lo largo de* (throughout), etc. A few adverbial phrases—for instance, *a pesar de* (in spite of), *de otra manera* (otherwise), etc.—and several pronominal phrases—such as *sí mismo* (itself)—are similarly treated. The list of such combinations is small (currently including 62 items) and closed. Though we currently do not perform named entity recognition, we plan to do so in the future.

4.3.1.2   Tagging

The text is part of speech (PoS)-tagged using the TnT tagger [25] trained on the Spanish corpus CLiC-TALP.[5] This tagger has a performance of over 94% [142].

We also correct some frequent errors of the TnT tagger. For example:

Rule	Example
Det Adj V → Det S V	*el inglés vino* "the English(man) came"
Det Adj Prep → Det S Prep	*el inglés con* "the English(man) with"

4.3.1.3   Lemmatizing

We use a dictionary-based Spanish morphological analyzer [86].[6] In case of ambiguity, the variant of the PoS that is reported by the tagger is selected, with the following exceptions:

Tagger predicted	Analyzer found	Example
Adjective	Past participle	*dado* "given"
Adverb	Present participle	*dando* "giving"
Noun	Infinitive	*dar* "to give"

If the analyzer does not give an option in the first column but does give one in the second column, the latter is accepted.

If an expected noun, adjective, or participle is not recognized by the analyzer, we remove a suffix, e.g., *flaquito* → *flaco* (little (and) skinny → skinny). To do so, we remove a suspected suffix and check whether the word is then recognized by the morphological analyzer. Examples of suffix removal rules include:

---

[5]clic.fil.ub.es

[6]www.Gelbukh.com/agme

Rule	Example
-cita → -za	*tacita → taza* "little cup → cup"
-quilla → -ca	*chiquilla → chica* "nice girl → girl"

4.3.2 Rules

Parsing rules are applied to the lemmatized text. Following an approach similar to that of [4, 34], we represent a rule as a sub-graph (e.g., N ← V) using the following steps:

1. A substring matching the sequence of words in the rule is searched for in the sentence.
2. Syntactic relationships between the matched words are established according to those specified in the rule.
3. All words that have been assigned a governor by the rule are removed from the sentence because they do not participate in further comparisons at step 1.

For example, for the sentence *Un perro grande ladra* (a big dog barks):

Sentence	Rule
Un(<u>Det</u>) *perro*(<u>N</u>) *grande*(Adj) *ladra* (V)	Det ← N
perro(<u>N</u>) *grande*(<u>Adj</u>) *ladra* (V) ↓ *Un*(Det)	N → Adj
perro(<u>N</u>) *ladra* (<u>V</u>) ↙ ↘ *Un*(Det) *grande*(Adj)	N ← V
ladra (V) ↓ *perro*(N) ↙ ↘ *Un*(Det) *grande*(Adj)	Done

As can be seen from that example, the order of the rule application is important. The rules are ordered; at each iteration of the algorithm, the first applicable rule is applied, after which the algorithm continues looking for subsequent applicable rules. Processing stops when no rule can be applied.

Note that one consequence of such an algorithm is its natural treatment of repeated modifiers. For example, in the phrases *el otro día* (the other day) and *libro nuevo interesante* (new, interesting book) the two determiners (both of which are adjectives) will be connected as modifiers to their respective nouns by the same rule Det ← N (N → Adj) at two successive iterations of the algorithm.

Our rules are not yet fully formalized (which is why we call our approach "semi-heuristic"), so in what follows we will provide additional comments to some rules. At present, the following rules are included[7]:

Rule	Example
Auxiliary verb system and verb chains	
estar | *andar* ← Ger	*estar comiendo* "to be eating"
haber | *ser* ← Part	*haber comido* "to have eaten"
haber ← *estado* ← Ger	*haber estado comiendo* "have been eating"
ir_{pres} *a* ← Inf	*ir a comer* "to be going to eat"
ir_{pres} ← Ger ← Inf	*ir queriendo comer* "keep wanting to eat"
V → *que* → Inf	*tener que comer* "to have to eat"
V → V	*querer comer* "to want to eat"
Standard constructions	
Adv ← Adj	*muy alegre* "very happy"
Det ← N	*un hombre* "a man"
N → Adj	*hombre alto* "tall man"
Adj ← N	*gran hombre* "great man"
V → Adv	*venir tarde* "come late"
Adv ← V	*perfectamente entender* "understand perfectly"
Conjunctions (see explanation below)	
N Conj N V(pl) ⇒ [N N] V(pl)	*Juan y María hablan* "John and Mary speak"
X Conj X ⇒ [X X] (X stands for any)	(*libro*) *nuevo e interesante* "new and interesting (book)"
Other rules	
N → *que* V	*hombre que habla* "man that speaks"
que → V	*que habla* "that speaks"
ⱱN X\que (X stands for any)	*hombre tal que* "a man such that"; *hombre, que* "man, which"
Det ← Pron	*otro yo* "another I"
V → Adj	*sentir triste* "to feel sad"
ⱱN, A\dj	*hombre, alto* "man, tall"
ⱱN, N\	*hombre, mujer* "man, woman"
N → Prep → V	*obligación de hablar* "obligation to speak"
ⱱV, V\	*comer, dormir* "eat, sleep"
V Det ← V	*aborrecer el hacer* "hate doing"

[7]The bar \| stands for variants: *estar* \| *andar* ← Ger stands for two rules, *estar* ← Ger and *andar* ← Ger.

Coordinative conjunctions have always been tricky for dependency formalisms and an argument in favor of constituency approaches. Following Gladki's idea [92], we represent coordinated words in a constituency-like manner, joining them in a compound quasi word. In the resulting "tree," we effectively duplicate (or multiply) each arc coming to or going from such a special node. For example, a fragment [*John Mary*] ← *speak* (*John and Mary speak*) is interpreted as representing two relationships: *John* ← *speak* and *Mary* ← *speak*. A fragment *merry* ← [*John Mary*] ← *marry* (*Merry John and Mary marry*) yields the following dependency pairs: *merry* ← *John* ← *marry* and *merry* ← *Mary* ← *marry*. We should note that, currently, this machinery is not yet fully implemented in our system.

Accordingly, our rules for handling conjunctions are more like rewriting rules than tree construction rules. The first rule forms a compound quasi word out of two coordinated nouns if they precede a plural verb. That rule eliminates the conjunction since, in our implementation, conjunctions do not participate in the tree structure. Basically, what the rule does is assure that the verb having such a compound subject is plural, i.e., it rules out the interpretation of *John loves Mary and Jack loves Jill* as *John loves* [*Mary and Jack*] *loves Jill*.

4.3.3   Prepositional Phrase Attachment

This stage is performed after the application of the rules described in the previous section.

For any preposition that has not yet been attached to a governor, its compatibility with every noun and every verb in the sentence is evaluated using word co-occurrence statistics (which can be obtained by a simple query to an Internet search engine). The obtained measure is combined with a penalty on the linear distance: the more distant a potential governor is from the preposition in question, the less appropriate it is for attachment. This will be discussed in the next chapter.

4.3.4   Heuristics

Heuristics are applied after the stages described in the previous sections. The purpose of the heuristics is to attach words that were not assigned to any governor during the rule application stage.

The system currently uses the following heuristics, which are iteratively applied in this order, in a manner similar to how the rules are applied:

1. An unattached *que* (that, which) is attached to the nearest verb (to the left or right of *que*) that does not have another *que* as its immediate or indirect governor.

2. An unattached pronoun is attached to the nearest verb that does not have *que* as
   its immediate or indirect governor.
3. An unattached N is attached to the most probable verb that does not have *que* as
   its immediate or indirect governor. To estimate probability, an algorithm similar
   to the one described in the previous section is used. The statistics described in
   (Calvo et al. 2005) are used.
4. For an unattached verb *v*, the nearest other verb *w* is looked for to the left; if
   there is no verb to the left, then the nearest one to the right is looked for. If *w* has
   que as a direct or indirect governor, then *v* is attached to that *que*; otherwise, it is
   attached to *w*.
5. An unattached adverb or subordinative conjunction (except for *que*) is attached
   to the nearest verb (to the left or right of *que*) that does not have another *que* as
   its immediate or indirect governor.

Note that if the sentence contains more than one verb, each verb will be attached
to another verb at step 4, which can result in a circular dependency. However, this is
harmless since such a circular dependency will be broken at the last stage of
processing.

4.3.5   Selection of the Root

The structure built during the algorithm's steps described in the previous sections
can be redundant. In particular, it can contain circular dependencies between verbs.
The final step of analysis, then, is to select the most appropriate root.

We use the following simple heuristics to select the root. For each node in the
obtained digraph, we count the number of other nodes reachable from the given one
through a directed path along the arrows. The word that maximizes this number is
selected as the root; in particular, all its incoming arcs are deleted from the final
structure.

Chapter 5
Evaluation of the Dependency Parser

Many corpora are annotated using constituent formalism. However, our goal is to evaluate parsers within the dependency formalism, which means we need a gold standard in the dependency formalism. In order to achieve this, we present an unsupervised heuristic (see Sect. 5.1) that aims to convert a constituent-annotated corpus into a dependency-annotated corpus, which then makes it possible to evaluate the parser presented in Chap. 4. A description of that evaluation can be found in Sect. 5.2.

5.1 Definition of a Gold Standard

Here, we present a method for converting an existing, manually tagged, constituent corpus into a dependency corpus. Roughly, this method consists of extracting a context-free grammar for the labeled text in order to automatically identify the head in each rule and then using that information to build a dependency tree. Our heuristics identify rules' heads with a precision of 99% and a coverage of 80%; the algorithm correctly identifies 92% of the dependency relationships between words in a sentence.

This section is organized as follows. Section 5.1.1 briefly introduces the constituent corpus that was the base for our experiments. Sections 5.1.2 and 5.1.3 present, in detail, the procedure we used to transform that corpus into a dependency corpus as well as the heuristics we used in that conversion. Section 5.1.4 discusses our experimental results, and Sect. 5.1.5 presents the conclusions.

5.1.1 The Spanish 3LB Treebank

Cast3LB is a corpus of one hundred thousand words (approximately 3,700 sentences) that was created from two other corpora: the CLiCTALP corpus (75,000 words), which is a balanced and morphologically annotated corpus containing

© Springer International Publishing AG 2018
A. Gelbukh and H. Calvo, *Automatic Syntactic Analysis Based on Selectional Preferences*, Studies in Computational Intelligence 765, https://doi.org/10.1007/978-3-319-74054-6_5

literary, journalistic, scientific, etc. language; and the corpus of the EFE Spanish news agency (25,000 words) that corresponds to the year 2000.

The annotation process was completed in two steps. In the first step, a subset of the corpus was selected and annotated twice by two different annotators; the results of this double annotation process were compared and a disagreement typology regarding assignation was established. After a process of analysis and discussion, an annotation handbook was produced, in which the main criteria to follow in cases of ambiguity are described. In the second step, the rest of the corpus was annotated according to the all-words strategy. The lexical items that are annotated are those words that have lexical meaning, i.e., nouns, verbs, and adjectives [143].

5.1.2 Transformation Procedure

The transformation procedure can be described roughly as follows:

1. Extract the constituency grammar rules from the 3LB treebank.
2. Use heuristics to find the head component of each rule.
3. Recursively use information regarding the heads to determine which rules will find which component to rise in the tree.

These steps are described in detail in the following sections.

5.1.3 Extracting the Grammar

To extract the grammar from the 3LB treebank, we used the following steps.

Simplification of the constituency treebank. The 3LB treebank divides tags into two parts. The first specifies the part of speech—for example, clause, noun, verb, noun phrase, etc. For our purposes, this is the most important part of the tag. The second part specifies additional features, such as gender and number for noun phrases or the kind of subordinate clause used. These features can be elided to reduce the number of grammatical rules included without affecting the transformation. For example, for a clause, the 3LB treebank uses: S (clause), S.F.C. (co-ordinate clause), or S.F.C.co-CD (object coordinate clause). We mapped each of these to a single label, *S*. For nominal groups, 3LB uses grup.nom (nominal group), grup.nom.fp (feminine plural nominal group), grup.nom.ms (masculine singular nominal group), grup.nom.co (coordinate nominal group), etc.; we mapped these to a single label, grupnom. Figure 5.1 shows part of the 3LB treebank using its original labels; Fig. 5.2 shows the same part using the transformed labels.

In order to reduce the number of the resulting grammar's patterns, we also simplified the tagging of the 3LB treebank by eliminating all punctuation marks.

(S	clause
(S.F.C.co-CD	clause
(S.F.C	clause
(sn-SUJ	noun phrase
(espec.fp	specifier
(da0fp0 Las el))	determiner *The* feminine plural *the*
(grup.nom.fp	nominal group
(ncfp000 reservas reserva)	noun *reserves reserve*
(sp	prepositional phrase
(prep	preposition
(sps00 de de))	preposition *of of*
(sn	noun phrase
(grup.nom.co	nominal group
(grup.nom.ms	nominal group
(ncms000 oro oro))	noun *gold gold*
(coord	coordinate
(cc y y))	coordinate *and and*
(grup.nom.fp	nominal group
(ncfp000 divisas divisa)))))	noun *currencies currency*
(sp	prepositional phrase
(prep	preposition
(sps00 de de))	preposition *from from*
(sn	noun phrase
(grup.nom	nominal group
(np00000 Rusia Rusia))))))	noun *Russia Russia*
(gv	verb phrase
(vmis3p0 subieron subir))	verb *raised to_raise*
(sn-CC	noun phrase
(grup.nom	nominal group
(Zm 800_millones_de_dolares	number *800_millions_of_dollars*
800_millones_de_dolares))))	*800_millions_of_dollars*

Fig. 5.1 A sentence with its original labels from the 3LB treebank. "The reserves of gold and currency from Russia rose by 800 million dollars"

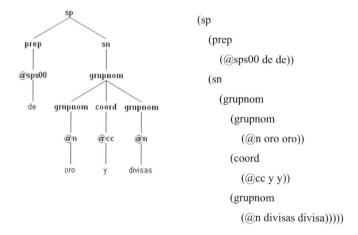

```
(sp
  (prep
    (@sps00 de de))
  (sn
    (grupnom
      (grupnom
        (@n oro oro))
      (coord
        (@cc y y))
      (grupnom
        (@n divisas divisa)))))
```

Fig. 5.2 Nodes that only have one leaf marked as the head

Pattern extraction. To extract all of the grammar's rules, each node with more than one child is considered to be the left part of a rule while their children are the right part of the rule. For example, the patterns extracted from the sentence shown in Fig. 5.3 are shown in Fig. 5.4. Here *grupnom* is a nominal group, *coord* is a

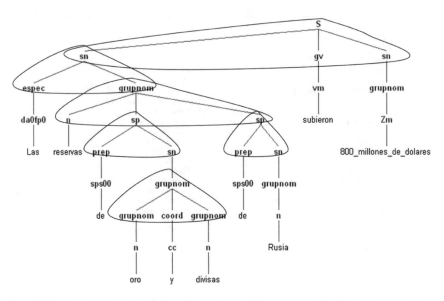

Fig. 5.3 Patterns to be extracted from the sentence. "The reserves of gold and currency from Russia rose by 800 million dollars"

Fig. 5.4 Extracted patterns of the sentence. "The reserves of gold and currency from Russia rose in 800 millions of dollars"	grupnom ← grupnom coord grupnom sp ← prep sn grupnom ← n sp sp sn ← espec grupnom S ← sn gv sn

coordinate, *sp* is a PP, *prep* is a preposition, *sn* is a noun phrase, *n* is a noun, *espec* is a specifier, *S* is a clause, and *gv* is a verb phrase. A clause (S) can be composed by a noun phrase (sn), verb phrase (gv), and noun phrase (sn).

5.1.3.1 Marking the Head

After extracting all patterns that form the grammar, the head of each pattern is automatically marked using simple heuristics. We denote the dead of a rule with the @ symbol. The heuristics we use are as follows:

1. If the rule contains only one element (or only one of its elements can be a head, see below heuristics 10 and 11), then it is the head, e.g.,

   grupnom ← @n

2. If the pattern contains one coordinate (*coord*), then it is the head, e.g.,

   grupnom ← grupnom @coord grupnom
   S ← @coord sn gv sn

3. If the pattern contains two or more coordinates, then the first is the head, e.g.,

   S ← @coord S coord S
   Sp ← @coord sp coord sp

4. If the pattern contains a verb phrase (*gv*), then it is the head, e.g.,

   S ← sn @gv sn
   S ← sadv sn @gv S Fp

5. If the pattern contains a relative pronoun (*relatiu*), then it is the head, e.g.,

   sp ← prep @relatiu
   sn ← @relatiu grupnom

6. If the pattern contains a preposition (*prep*) as its first element, followed by only one element (regardless of what that element may be), then the preposition is the head, e.g.,

   sp ← @prep sn
   sp ← @prep sp

7. If the pattern contains an infinitive verb (*infinitiu*), then it is the head, e.g.,

   S ← @infinitiu S sn
   S ← conj @infinitiu
   S ← neg @infinitiu sa

8. If the pattern contains a present participle (*gerundi*), then it is the head, e.g.,

   S ← @gerundi S

9. If the pattern contains a main verb (*vm*), then it is the head, e.g.,

   gv ← va @vm
   infinitiu ← va @vm

10. If the pattern contains an auxiliary verb (*va*) and any other verb, then the auxiliary verb is never the head, e.g.,

    gv ← va @vs

11. If the pattern contains a specifier (*espec*) as its first element, then it is never the head, e.g.,

    sn ← espec @grupnom
    sn ← espec @sp

12. For patterns with a noun phrase (*grupnom*) as the father node, if the pattern contains a noun (*n*), then it is the head, e.g.,

    grupnom ← s @n sp
    grupnom ← @n sn
    grupnom ← s @n S

13. For patterns with a noun phrase (*grupnom*) as the father node, if the pattern contains a noun phrase (*grupnom*), then it is the head, e.g.,

    grupnom ← @grupnom s
    grupnom ← @grupnom sn

14. For patterns with a specifier (*espec*) as the father node, if the pattern contains a definitive article (*da*), then it is the head, e.g.,

    espec ← @da di
    espec ← @da dn

15. If the pattern contains a qualificative adjective (*aq*) and a PP (*sp*), then the adjective is the head, e.g.,

    S ← sadv @aq sadv
    sa ← sadv @aq sp sp

The application order of these rules is important. For example, if we apply rule 2 before rule 1 in the pattern—S ← coord sn gv sn Fp—the head would be *gv* instead of the correct head *coord*. Hence, rule 1 should be applied first.

5.1.3.2   Using Marked Heads for the Transformation

The transformation algorithm recursively uses the information of the patterns marked with heads to determine which components will rise in the tree. This means that the head will be disconnected from its brothers and placed in the father node position.

In order to more clearly understand the algorithm, we describe it in detail:

1. Traverse the constituency tree in depth from left to right, beginning at the root and visiting the children nodes recursively.
2. For each pattern in the tree, search the rules to find which element is the head.
3. Mark the head in the constituency tree. Disconnect it from its brothers and place it in the father node position.

The algorithm finishes when a head node is raised as a root. For example, consider the following figures.

Figure 5.5 shows a constituency tree that will be transformed into a dependency tree. Remember that nodes with only one leaf were marked in the extraction grammar.

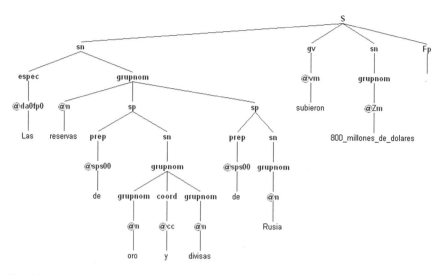

Fig. 5.5 Constituency tree. "The reserves of gold and currency from Russia rose by 800 million dollars"

After the algorithm, the first pattern to be found is: grupnom ⟵ grupnom coord grupnom, where *grupnom* is a nominal group and *coord* is a coordinate.

Looking at the rules we find that the head of these patterns is the coordinate (coord). We mark the head in the constituency tree and disconnect it by putting it in the father node position, as shown in Fig. 5.6.

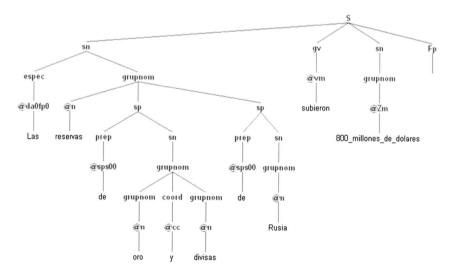

Fig. 5.6 Constituency tree. "The reserves of gold and currency from Russia rose by 800 million dollars"

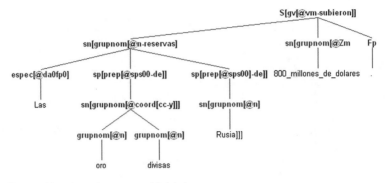

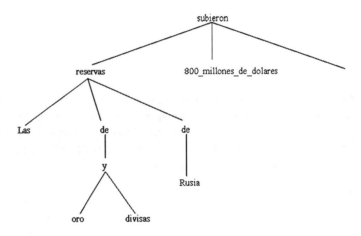

Fig. 5.7 Resulting dependency tree with labels

Fig. 5.8 Resulting dependency tree without labels

The algorithm completes its execution until the root node is raised. The resulting dependency tree is shown in Figs. 5.7 and 5.8.

5.1.4   Experimental Results

The algorithm found 2663 grammar rules. From those, 339 (12%) are repeated more than 10 times and 2324 (88%) less than 10 times. The twenty most frequent rules (with their respective number of occurrences) are:

12,403 sn ← espec grupnom
11,192 sp ← prep sn
3229 grupnom ← n sp

1879 grupnom ← n s
1054 sp ← prep S
968 grupnom ← n S
542 gv ← va vm
535 grupnom ← s n
515 S ← infinitiu sn
454 grupnom ← n s sp
392 grupnom ← n sn
390 grupnom ← grupnom coord grupnom
386 sn ← sn coord sn
368 grupnom ← s n sp
356 gv ← vm infinitiu
343 S ← S coord S Fp
315 S ← S coord S
276 sp ← prep sn Fc
270 grupnom ← n sp sp
268 S ← infinitiu sp

5.1.4.1 Head Identification

The heuristics covered (i.e., automatically labeled) 2210 (79.2%) of all extracted grammar rules.

We randomly selected 300 rules and marked them manually. Comparison showed that all but two (99.9%) marks coincided (see Fig. 5.9). These two rules are not matched because the heuristic rules do not consider these cases.

Based on these comparison statistics, we believe that at least 95% of the automatically marked rules from 3LB are correctly marked.

5.1.4.2 Construction of Dependency Trees

We followed the evaluation scheme proposed by Briscoe et al. [29], in which parsing accuracy is evaluated based on grammatical relationships between lemmatized lexical heads. This scheme is suitable for evaluating dependency and constituency parsers because it considers tree relationships that are present in both formalisms—for example, [Det *car the*] and [DirectObject *drop it*]. For our evaluation, we extracted triples from the dependency trees obtained by our method and compared them with manually extracted triples from the same 3LB treebank.

Automatically marked	Manually marked
infinitiu <-- van0000 vmp00sm sps00 @infinitiu	infinitiu <-- van0000 @vmp00sm sps00 infinitiu
S.F.C.co-CD <-- conj.subord S.F.C @coord S.F.C	S.F.C.co-CD <-- @conj.subord S.F.C coord S.F.C

Fig. 5.9 Rules that did not match

A triple is a dependency relationship between a father node, a child node, and the type of their relationship. For example, the dependency triples extracted from the phrase *The old man loves the young lady* are:

love SUBJ man
man DET the
man ADJ old
love OBJ lady
lady DET the
lady ADJ young

The algorithm extracted 65,997 dependency triples from the whole 3LB treebank.

For evaluation, we randomly selected 35 sentences from the treebank and manually converted them to dependency trees, which gave us 399 dependency triples. We then applied our procedure to those sentences. Since, for a sentence of n words, there must be $(n - 1)$ triples, our procedure also output 399 triples; of them, 368 (92%) coincided with those manually identified. Extrapolating this statistic, we infer that more than 90% (some 60,000) of the dependency triples that we extracted from the 3LB treebank are correct.

5.1.5   Conclusions

Dependency representation of syntactic structures has important advantages in certain applications—particularly in regards to nearly everything related to lexicalization and lexicography. However, the majority of existing tools and resources, such as parsers, grammars, and treebanks, are oriented to a constituency approach.

We have presented a simple, unsupervised technique that allows for the automatic transformation of constituency trees into dependency trees. This technique uses certain simple heuristics that depend on the specific tag set used in a given treebank or grammar. Our technique does not deal with difficult or arguable phenomena in dependency syntax but still recovers the bulk of dependency relations. Such "quick-and-dirty" results are quite usable in most practical applications.

Thus, our technique allows for existing parsers or treebanks to be reused for applications requiring dependency structures.

5.2   Evaluation of Our Parser

This section presents a comparison of our parser against a hand-tagged gold standard. We also compared our parser with two widely known Spanish-language parsers: Connexor Machinese Syntax for Spanish (a dependency parser) and TACAT (a constituency parser).

We followed the evaluation scheme proposed by Briscoe et al. [29], which evaluates parsing accuracy based on grammatical relationships between lemmatized lexical heads. This scheme is suitable for evaluating both dependency and constituency parsers because it considers tree relationships that are present in both formalisms—for example, [Det *car the*] and [DirectObject *drop it*]. For our evaluative purposes, we translated the output of the three parsers and the gold standard into a series of triples that included two words and their relationship. The triples from the parsers were then compared against the triples from the gold standard to find correspondence.

We chose the corpus Cast3LB as our gold standard because it is, until now, the only syntactically tagged corpus for Spanish that is widely available. Cast3LB is a corpus consisting of 100,000 words (approximately 3,700 sentences) extracted from two corpora: the CLiCTALP corpus (75,000 words), a balanced corpus containing literary, journalistic, scientific, and other topics; and the EFE Spanish news agency (25,000 words) corpus that corresponds to the year 2000. That corpus was annotated according to Civit and Martí [53] using the constituency approach; so we first convert it to a dependency treebank. A rough description of this procedure follows. For details, see Gelbukh et al. [88].

1. Extract patterns from the treebank to form rules. For example, a node called NP with two children, Det and N, yields the rule NP → Det N.
2. Use heuristics to find the head component of each rule. For example, a noun will always be the head in a rule except when a verb is present. The head is marked with the @ symbol: NP → Det @N.
3. Use this information to establish the connections between heads of each constituent.
4. Extract triples for each dependency relationship in the dependency treebank.

For example, consider the following table, which shows the triples for the sentence taken from Cast3LB. *El más reciente caso de caridad burocratizada es el de los bosnios, niños, y adultos* ("the most recent case of bureaucratized charity is the one about Bosnians, children, and adults"). In some cases, the parsers extracted additional triples not found in the gold standard.

Spanish triples	Gloss	3LB	Connexor	DILUCT	TACAT
adulto DET el	'the adult'	✓			
bosnio DET el	'the bosnian'	✓	✓	✓	
caridad ADJ burocratizado	'bureaucratized charity'	✓		✓	✓
caso ADJ reciente	'recent case'	✓		✓	✓
caso DET el	'the case'	✓		✓	✓
caso PREP de	'case of'	✓	✓	✓	✓
de DET el	'of the'	✓			✓
de SUST adulto	'of adult'	✓			

(continued)

(continued)

Spanish triples	Gloss	3LB	Connexor	DILUCT	TACAT
de SUST bosnio	'of bosnian'	✓		✓	
de SUST caridad	'of charity'	✓	✓	✓	✓
de SUST niño	'of children'	✓			
niño DET el	'the child'	✓			
reciente ADV más	'most recent'	✓			✓
ser PREP de	'be of'	✓		✓	✓
ser SUST caso	'be case'	✓		✓	✓
recentar SUST caso	'*to recent* case'		✓		
caso ADJ más	'case most'			✓	
bosnio SUST niño	'bosnian child'			✓	
ser SUST adulto	'be adult'			✓	
de,	'of,'				✓
, los	', the'				✓
, bosnios	', Bosnian'				✓

We extracted 190 random sentences from the 3LB treebank and parsed them using Connexor and DILUCT. The precision, recall, and F-measure of the different parsers against Cast3LB are as follows:

	Precision	Recall	F-measure
Connexor	0.55	0.38	0.45
DILUCT	0.47	0.55	0.51
TACAT[1]	–	0.30	–

Note that the Connexor parser, although it has a slightly better precision and a rather similar F-measure as our system, is not freely available and is, of course, not an open source.

5.3   Conclusions

We have presented a simple and robust dependency parser for Spanish. It uses simple handmade heuristic rules for decisions regarding the admissibility of structural elements and word co-occurrence statistics for disambiguation. The statistics are either learned from a large corpus or obtained by querying an Internet search engine in an unsupervised manner—i.e., no manually created treebank is used for training. In case the parser cannot produce a complete tree, a partial structure is returned that consists of the dependency links it could recognize.

---

[1]Results for TACAT were kindly provided by Jordi Atserias.

For details of our evaluation of this system, see Sect. 5.2. A comparison of the accuracy of our parser with two available systems for Spanish shows that our parser outperforms both.

Though a number of grammar rules are specific for Spanish, the approach itself is language-independent. In the future, we plan to develop similar parsers for other languages, including English, for which the necessary preprocessing tools—such as a PoS tagger and lemmatizer—are already available.

One possible direction our future work may take is improving the grammatical rules system. The current rules sometimes do their job in a quick-and-dirty manner, which—while resulting in just the right thing to do in most cases—can be done with greater attention to details.

Finally, we plan to evaluate the usefulness of our parser for real information retrieval, text mining, and construction of semantic textual representations (such as conceptual graphs).

Chapter 6
Applications

6.1 Selectional Preferences

In this chapter, we propose a method for extracting selectional preferences that are linked to an ontology: namely, WordNet. This information is used, among other possible applications, to perform word sense disambiguation. An evaluation of this method, using Senseval-2, is also given. The results of this experiment are comparable to those obtained by Resnik when he used selectional preferences for the English language; however, our proposed method is more advantageous in that it does not require any previous morphological, syntactic, or semantic annotation of the text.

6.1.1 Introduction

Selectional preferences measure an argument's coupling degree (direct and indirect objects as well as prepositional complements) with regard to a verb. For example, for the verb *to drink*, the direct objects *water, juice, vodka*, and *milk* are more probable than *bread, ideas*, or *herb*.

For a system to have as large a potential coverage of verb complements as possible, it is necessary to have a very big training corpus. However, even very big corpora of hundreds of millions of words fail to include certain word combinations —even those that may be of everyday use.

One possible solution to this problem is to use word classes. In this case, *water, juice, vodka*, and *milk* belong to the *liquid* class; thus, we can establish an association between the *liquid* class and the verb *to drink*. However, not all verbs have this kind of specific association. For example, the verb *to take* could have arguments from many different classes: *to take account, to take seat, to take advantage...* .

© Springer International Publishing AG 2018
A. Gelbukh and H. Calvo, *Automatic Syntactic Analysis Based on Selectional Preferences*, Studies in Computational Intelligence 765,
https://doi.org/10.1007/978-3-319-74054-6_6

On the other hand, each word can have more than one classification—in which case, both the meaning (or sense) of a word and its characteristics must be considered to determine whether it belongs to a specific class. For example, if we consider the color of objects, we would select the *white objects* class for milk. If we consider physical properties, we would say that *milk* belongs to the *fluids, liquids* class or even to the *antacids* class. The selected class for *milk* could also be *basic_food*, and so forth. In other words, the relevant classification for a word depends on its intended use—not only on its meaning.

In order to find relationships between the use and sense of a noun and the selectional preferences of a verb, two kinds of information are necessary: (1) the ontological information of a word so that the word will not be linked in a flat manner to a single class, and (2) information about the use of a word, given a verb, linked to a specific position in the ontology.

In the following section, we describe a method to extract selectional preferences that are linked to an ontology. This information helps to solve several problems within the area of statistical models combined with knowledge [165, 166]. For example, refer to Fig. 6.2, which gives arguments for three verbs using the WordNet ontology hierarchy. The aim of our method is to obtain a similar table and, subsequently, use that information to perform word sense disambiguation (WSD).

6.1.2 Related Works

One of the first works on selectional preference extraction linked to WordNet senses is Resnik's [167], which is devoted mainly to word sense disambiguation in English. Resnik assumed that a text annotated with word senses was a difficult-to-obtain resource, so he based his work on text that was tagged only morphologically. Subsequently, Agirre and Martinez [1, 2] linked verb usage with their arguments. In contrast with Resnik, Agirre and Martínez assumed the existence of text that is annotated with word senses: Sem-Cor, in English. Other supervised WSD systems include John Hopkins University's system (JHU) [208], which won the Senseval-2 competition, and a maximum entropy WSD system by Suarez and Palomar [187]. The first system combined, by means of a voting-based classifier, several WSD subsystems that were based on different methods: decision lists [207], cosine-based vector models, and Bayesian classifiers. The second system selected a best-feature selection for classifying word senses and a voting system. These systems both had scores of around 0.70 in the Senseval-2 tests.

We take into account the fact that a resource like Sem-Cor is not currently available for many languages and that the cost of building such a resource is high. Accordingly, we follow Resnik's approach in assuming that there is not a large enough quantity of text annotated with word senses. Furthermore, we consider that the WSD process must be completely automatic, so all the text we use is automatically tagged with morphological and part-of-speech (PoS) tags. Accordingly, our system is fully unsupervised.

Previously, unsupervised systems have not achieved the same performance as that achieved by supervised systems: Carroll and McCarthy [45] present a system that uses selectional preferences for WSD that obtains 69.1% precision and 20.5% recall; the unsupervised method presented by Agirre and Martinez [3] obtains 49.8% recall, which is the only performance measure used; Resnik's unsupervised method [167] achieves 40% correct disambiguation.

The following sections describe our method and measure its performance.

6.1.3 Methodology

In order to enhance an ontology with usage values for its synsets, we used Spanish EuroWordNet 1.0.7 (S-EWN)[1] and a corpus of 161 million words that correspond to four years' worth of publications from three separate Mexican newspapers. This corpus contains approximately 60 million sentences. The text was automatically PoS tagged using the statistical PoS tagger TnT, which was trained with the CliC-TALP corpus. We found, as well as [142], that this tagger has about 94% accuracy for Spanish. Afterward, simple grouping rules were used to find chunks of adjective+noun, adverb+verb, etc. Subordinate phrases were included.

Once the text was tagged, the following combinations were extracted: verb +left_subject, verb+right_subject, and verb+preposition+noun. Here the symbol "+" indicates immediate adjacency. We avoided patterns that included intermediate words in order to avoid possible noise; for example, patterns such as verb+X +preposition+noun (in which X replaces any word or set of words) were discarded.

Figure 6.1 provides an example of word patterns extracted in this way. The symbol ">" indicates that the noun appears immediately to the right of the verb (thus, becoming an object), while the symbol "<" indicates that the noun appears immediately to the left of the verb (thus, is a subject). A preposition corresponds to the form verb+preposition+noun.

Afterward, the noun of each combination is looked up in WordNet and its frequency increased by one. In the case that this noun has more than one sense, the frequency is evenly distributed among each of those senses. For example, if *book* has five senses, each one receives a count of 1/5.

This value is propagated up the WordNet hierarchy by following the hypernyms of each sense. The higher the level, the less impact a frequency count should have; thus, this value is multiplied by $\frac{1}{level}$. For example, for the first lower level (the leaf), the value for the occurrence of a verb for that synset is increased by 1; in the second level, it is increased by 0.5; in the third, by 0.33; etc. In the first example shown in

---

[1]S-EWN was jointly developed by the University of Barcelona (UB), Universidad Nacional de Educación a Distancia (UNED), and Universidad Politécnica de Cataluña (UPC), Spain.

have a permission:
00629673n → **sanction** 58.96 → authorization 231.89 → management 83.01 → social_control 115.54 → action 1808.92

04368291n → **approval** 232.57 → message 570.55 → communication 1066.30 → social relationship 761.52 → relation 847.98 → abstraction 1734.82

08562692n → **liberty** 198.76 → freedom 198.76 → state 640.99

cross > channel
02342911n → **way** 3.00 → trough 8.83 → artifact 20.12 → unanimated_obect 37.10 → entity 37.63

02233055n → **conduit** 6.00 → way 3.00 → trough 8.83 → artifact 20.12 → unanimated_object 37.10 → entity 37.63

03623897n → **conduit** 5.00 → anatomic_structure 5.00 → body_part 8.90 → part 7.22 → entity 37.63

04143847n → **transmission** 1.67 → communication 3.95 → action 6.29

05680706n → **depression** 2.33 → geological_formation 2.83 → natural_object 14.50 → unanimated_object 37.10 → entity 37.63

05729203n → **water** 4.17 → unanimated_object 37.10 → entity 37.63

read > book:
01712031n → **(tripe) stomach** 51.30 → internal_organ 34.20 → organ 29.01 → body_part 31.90 → piece 53.86 → entity 271.28

02174965n → **product** 177.33 → creation 164.83 → artifact 209.80 → inanimate_object 232.52 → entity 271.28

04214018n → **section** 93.82 → writing 514.13 → written_language 377.39 → communication 831.37 → social_relationship 645.04 → relation 628.82 → abstraction 587.39

04222100n → **publication** 106.00 → work 282.98 → product 177.33 → creation 164.83 → artifact 209.80 → inanimate_object 232.52 → entity 271.28

04545280n → **dramatic_work** 74.49 → writing 514.13 → written_language 377.39 → communication 831.37 → social_relationship 645.04 → relation 628.82 → abstraction 587.39

Fig. 6.1 Ontology with usage values for the combinations *cross channel, have a permission,* and *read book*

Fig. 6.2, *permission* has three different senses. Figure 6.3 shows the WordNet fragment corresponding to *permission*. When the combination *have a permission* is seen in the corpus used for counting occurrences, one count will be added for *permission* (*have a*); $\frac{1}{2} \cdot \frac{1}{3}$ will be added for the sense of *permission* that corresponds to *sanction, approval*, and *liberty*; $\frac{1}{3} \cdot \frac{1}{3}$ will be added for its sense that corresponds to *authorization, message, and freedom*; and so on.

These values were added for each occurrence in the counting corpus. The result was an ontology with weighted usage numbers (selectional preferences) for each verb's arguments. Figure 6.1 shows the results for the combinations *have a permission* and *read* > book; numbers appearing in that figure correspond to the corresponding synset correspondence in E-WordNet.

Fig. 6.2 Sample combinations

1	have	a	permission
2	retire	from	system
3	hit	>	ball
4	solve	>	problem
5	give	>	signal
6	existe	<	incognite
7	put	in	pan
8	take	from	source
9	drink	>	vodka

Fig. 6.3 WordNet structure
for *permission*

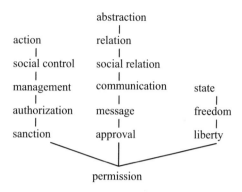

Figure 6.1 shows that this information can be used to choose the most probable sense of a word, since we know the verb related to that word. For example, for *read*, the less frequent sense is *stomach* (*tripe*); the most probable one is *product* (*creation*), followed by *publication* (*work*). *Channel* has six senses listed by WordNet: *way, conduit, clear, conduit* (anatomic), *transmission, depression*, and *water*. The sense marked with the highest number of occurrences is *conduit*, while the one with fewer occurrences is *transmission* (meaning, for example, the *channel of transmission* or *TV channel*); one cannot *cross* a TV channel. Currently, we have acquired 1.5 million selectional preference patterns that are linked to the WordNet synsets. Each pattern consists of a verb, a preposition (in some cases), and a synset. This information can be used to disambiguate the sense of the word, given the verb with which it is used. In the next section, we describe an experiment in which we measured the WSD performance of this method.

6.1.4   Evaluation

Senseval is a series of WSD evaluation exercises organized by the ACL-SIGLEX (Special Interest Group on the Lexicon of the Association for Computational Linguistics). Data for these competitions are available online. We applied our method to the Spanish 2001 competition.

The evaluation set comprises slightly more than 1000 sentences. Each sentence contains one word, for which the correct sense, among those listed for it in WordNet, is indicated.

Our evaluation showed that 577 of 931 cases were resolved (a recall of ∼62%). Of those, 223 corresponded in a fine-grained way to the sense that was manually annotated (precision ∼38.5%). These results are similar to those obtained by Resnik [167] for English, who obtained, on average, 42.55% precision for relationships between subjects and verbs only. Note that these results are much better than those obtained through the random selection of senses (around 28% as reported in [167]).

6.1.4.1 Discussion

Results show a lower performance for our proposed system compared with supervised WSD systems—for example, Suarez and Palomar [187] report a score of 0.702 for the same evaluation set of nouns from Senseval-2. In comparison with existing unsupervised WSD systems (i.e., [3, 45, 167) our method has better recall but lower precision, in some cases, because our method only considers verb-noun relationships and sometimes a word's sense is strongly linked to a preceding noun (this is particularly true for pairs of nouns that form a single PP).

On the other hand, our method has relatively low coverage since we only consider verb-subject and verb-object relationships. Other WSD methods may rely on other relationships, such as adjective-noun and other relationships involving modifiers. For example, we find the following text in the evaluation set: *The Apocalypse has nothing to do with Star Wars or the atomic bomb*. The sense of the word *bomb* is completely determined by its adjective (*atomic*)—not by the central verb of the subordinate clause (*do*). Thus, in this case, determining the sense of *bomb* with the full combination of *do, with*, and *bomb* is not the best strategy.

In order to improve this method, we can include more information regarding combinations of adjectives and adverbs.

6.1.5 Other Applications

Besides WSD, information regarding the selectional preferences obtained by this method can be used to solve important problems such as syntactic disambiguation. For example, consider the phrase *Pintó un pintor un cuadro* (literally, "painted a painter a painting," which means "a painter painted a painting"). In Spanish, it is possible to put the subject to the right of the verb. In English, there is ambiguity since it is not possible to decide which noun is the subject of the sentence; however, because in Spanish a rather free word order is used, even *Pintó un cuadro un pintor* (literally, "painted a painting a painter") has the same meaning.

For languages in which such freedom of word order does not exist, it is possible consult the ontology linked with the selectional preferences already constructed using our proposed method in order to decide which word is the subject (*painting* or *painter*). First, we find that the subject appears to the left of the verb 72.6% of the time [139]. Then, searching for *un pintor pintó* ("a painter painted") returns the following chain of hypernyms with occurrence values: *painter* → *artist* 1.00 → *creator* 0.67 → *human_being* 2.48 → *cause* 1.98. Finally, the search of *un cuadro pintó* ("a painting painted") returns *scene* → *situation* 0.42 → *state* 0.34. That is, *painter* (1.00) is more likely to be the subject than *painting* (0.42) for this sentence. A large-scale implementation of this method is a planned topic of future work.

6.1.6   Conclusions

We have presented a method for extracting selectional preferences that are linked to an ontology and applied those preferences to WSD. The obtained information is useful for solving several tasks that require information about the use of words in relationship to a verb in a sentence. The results of our WSD evaluation show that, despite still having much room for improvement, we have obtained results that are comparable with those of previous works that assume no morphological, PoS, or semantic annotation.

6.2   Steganography

Linguistic steganography allows information to be hidden in a text. The resulting text must be grammatically correct and semantically coherent to be accurate. Among several methods of linguistic steganography, we adhere to previous approaches that use synonymous paraphrasing, i.e., substituting content words with their equivalent. Context must be considered in order to avoid substitutions that could disrupt the text's coherence (for example, *spicy dog* instead *hot dog*). We base our method on previous works in linguistic steganography that use collocations to verify context, but propose using selectional preferences instead of collocations because selectional preferences can be collected automatically in a reliable manner, thus allowing our method to be applied to any language.

Our work is based on previous works [15, 17] that use a manually collected collocations database. However, manually collecting collocations could take many years; one Russian collocations database [16] took more than 14 years to complete. On the other hand, using the Internet to verify collocations [20] is not adequate for split collocations (such as *make an* awful *mistake*) because current web search engines do not allow such searches—the closest search tool is the NEAR operator, which is not precise because it is not restricted to a single sentence.

6.2.1   Some Definitions

Linguistic steganography comprises a set of methods and techniques that allow information to be hidden in a text based on the reader's own linguistic knowledge. To be effective, the resulting text must have grammatical correctness and semantic cohesion.

There are two main approaches to linguistic steganography: (1) generating text and (2) changing previously written text. To illustrate the first approach, consider a verb–preposition–noun sentence model. This model can generate valid sentences such as *go to bed, sing a song*, etc. However, a non-trivial problem arises when trying to generate a coherent text using this sentence: *John goes to bed, and then John sings a song*. This type of non-coherent text is not free of suspicion. As Chapman et al. [46] indicates, the same happens when using more elaborate sentence models extracted from previously written text.

In the second approach, some words in the source text are replaced by other words, depending on the bit sequence to be hidden. These changes are detectable only at the intended receiver's side. In the best cases, the resulting text maintains the original meaning of the source text.

As in [15, 17, 20], we adhere to the second approach because it is far more realistic; indeed, generating text from scratch requires not only syntactic and semantic information but also pragmatic information.

In this work, we do not consider other methods of textual steganography such as text formatting, varying space width, or other non-linguistic encoding methods. We do not consider such methods because genuine linguistic methods allow a message to be transmitted independent of the medium—linguistic steganography allows a message to be transmitted over the internet, the telephone, a radio broadcast, etc.

To put it in other words, we continue to develop the method of linguistic steganography that replaces textual words by their synonyms [15]. This work allows information to remain concealed in unsuspicious texts along with linguistic correctness and the original meaning of the source text. In addition, we take advantage of existing resources to extend coverage of this method to virtually any language—whenever the resources required exist for that language. We do so by considering an alternative source for the context of words.

6.2.2   The Context of Words

The context of a given word is the words that surround it within a sentence. In many studies, only the surrounding words are considered when forming collocations with the given word. Other authors consider collocations as *a sequence of two or more consecutive words that has characteristics of a syntactic and semantic unit and whose exact and unambiguous meaning or connotation cannot be derived directly from the meaning or connotation of its components*, as defined by Choueka [51]. Examples that fall within this definition, including idioms, are *hot dog, white wine* (actually white wine is yellow) [122], *kick the bucket,* and *piece of cake.*

Furthermore, current usage of the term *collocation* also includes combinations of words that maintain their original meaning (such as *strong coffee*) but are considered collocations because substituting any one of their components with equivalent words yields an understandable but unusual combination (such as *powerful coffee, strong rain* (instead of *heavy rain*), and *do a mistake* (instead of *make a mistake*)).

Moreover, collocations are not necessarily adjacent words, e.g., *making an* awful *mistake* may involve subcategorization issues, as we illustrate below.

The links between components of collocations are syntagmatic. These are, for example, the link between a verb and a noun filling its valence (*made* → *of stone*) or the link between a noun and its adjective modifier (*black* ← *stone*). This type of relationship can be clearly seen in a dependency representation. The **context of a word** is given then by its dependency relationships. For example, consider the sentence *Mary read us a fairy tale* in Fig. 6.4.

To illustrate the influence of a word's context in a sentence, we substitute words with their equivalent (i.e., synonyms). For example, among synonyms of *fairy* are *pixie* and *nymph*. Substituting *fairy* by these equivalents yields, however, a very strange sounding sentence—*Mary read us a pixie tale* or *Mary read us a nymph tale*—which is possible but sounds odd since *fairy* depends strongly on *tale*. Another example is substituting *yarn* for *tale*—forgetting about *fairy* for a moment. In this case, it is odd to say *read us a yarn*, since it would be much more natural to say *spin us a yarn* instead—not considering *spin us a fairy yarn*! This shows the strength between the verb (*read* and *spin*) and one of its arguments (*tale* and *yarn*, respectively).

Subcategorization has an important role in collocation consideration. For example, consider the synonyms for *to tell: to relate, to spin,* and *to say*. If one wanted to change *read* for one of its synonyms, its context and structure must be considered in order to maintain the same meaning—and naturalness—of the sentence. Simply changing *read* to *related* would yield *Mary related us a fairy tale;* for this to be a natural-sounding sentence, a different structure (i.e., subcategorization frame) should be used: *Mary related a fairy tale to us.*

In contrast to this latter example, in this work we focus only on synonyms that maintain the structure and word order of a sentence as well as the number of words —counting stable multi-words such as *hot dog* as one unit. We use word combinations to verify that the synonymous paraphrasing results in a natural and coherent text.

6.2.3 Verifying Word Combinations

Our goal is to create synonym paraphrasing that considers context. In [15, 17] a previously collected collocation, DB was used. Currently, however, only a few

Fig. 6.4 Simplified dependency representation tree for *Mary read us a fairy tale*

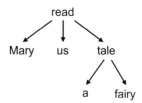

electronic collocation databases are readily available. To our knowledge, such publicly available databases did not exist until 1997, when the Advanced Reader's Collocation Searcher (ARCS) for English emerged [14]; however, this is now inferior to the Oxford Collocation Dictionary [148] in all respects.

The only project in the last decade to develop a very large collocation DB for local use was dedicated to the Russian language and produced an interactive system called CrossLexica [16, 18, 19]. It is mainly a large database of Russian collocations, but it also contains the equivalent of a Russian WordNet that contains a synonym dictionary and a hyponym/hypernym hierarchy.

A manually collected DB of collocations cannot list every possible pair of words —especially free word combinations such as *big boy*, *walk in the street*, etc. Thus, several methods for extracting collocations automatically are described in [122, 154]. However, the quality of these automatically obtained collocations is not as good as the quality of those obtained by hand. In addition, as we showed in Sect. 6.2.2, the context of a word is strongly related to the structure of the sentence. Hence, we need linguistic knowledge in addition to purely statistic methods.

Furthermore, polysemous words have several synonyms that cannot substitute the original word without changing the meaning of the text because these synonyms are for other senses of the word. For example *plant* can be substituted by *vegetable* or by *factory*, depending on context.

Till date, we have identified the following requirements for automatic determination of possible word combinations: a corpus (from which to learn), semantic knowledge, and sentence structure. The linguistic knowledge we use, which covers semantics and allows determination of sentence structure, is a set of **selectional preferences**.

6.2.4   Selectional Preferences for Synonym Paraphrasing

Selectional preferences measure the degree to which a verb *prefers* an argument—a subject, an object, or a circumstantial modifier. The selectional preferences principle can also be applied to adjective-noun relationships, verb-adverb relationships, and PPs, thus yielding a database of *preferences* that can be regarded as graded collocations with the aid of semantic generalizations. For example, if *harvest plants* appears in a training corpus and if we know that *harvest* prefers arguments of the type *flora*, then we can restrict synonyms for *plant* to those that are only related to *flora*, thus excluding those related to manufacturing processes.

In addition, selectional preferences can aid in the determination of sentence structures. For example, the syntactic structure of *I see the cat with a telescope* is disambiguated considering that *see with* {*an instrument*} is more frequently used than is *cat with a* {*instrument*}. Calvo and Gelbukh propose a method for PP attachment disambiguation in [41] and show how this information can be used to restrict a word's meaning in [38].

For this work, we use a selectional preferences database that is based on a corpus containing four years' worth of Mexican newspapers with 161 million words, as in [39].

6.2.5 The Algorithm

The proposed steganographic algorithm has two inputs:

- The information to be hidden, in the shape of a bit sequence.
- The source text, which must be written in a natural language with a minimal length that is approximately 500 times greater than that of the information to be hidden. The text format can be arbitrary, but the text proper should be orthographically correct so as to lessen the probability of unintended corrections during transmission. Such corrections can change the number of synonymous words in the text or the conditions for their verification and, thus, desynchronize the steganography versus steganalysis. The text should not be semantically specific, i.e., it should not be a mere list of names or sequence of numbers. In this respect, newswire flows or political articles are quite acceptable. Any long fragments of an inappropriate type increase the total length required for steganographic use.

The steps of the algorithm follow:

A1. Tagging and lemmatizing. The text is tagged using the TnT tagger, which has been trained with the Spanish corpus LEXESP [179]. That corpus has been reported to be 94% accurate for the Spanish language [142]. The text is then lemmatized by trying morphological variants against a dictionary [116].

A2. Identification of word combinations that can be paraphrased. The following patterns are extracted for each sentence; subordinate clauses are treated as separated sentences so that there is only one verb per sentence.

- (i)   noun+verb
- (ii) verb+noun
- (iii) noun+preposition+noun
- (iv) verb, preposition+noun

The symbol "+" denotes adjacency while a comma denotes *near to*—in the same sentence. All other words—adverbs, adjectives, articles, etc.—are discarded.

For patterns iii and iv, ambiguity is possible when certain nouns are attached to either the previous noun or the main verb of the sentence. For example, in *I eat rice with chopsticks*, the noun *chopsticks* could be attached to either *rice* or *eat*. This ambiguity is solved by considering the strengths of the selectional preferences of both possibilities. Only the strongest combination is considered [39]. In the previous example, *eat with chopsticks* (pattern iv) is stronger than *rice with chopsticks* (pattern iii). Conversely, in *I eat rice with beans*, *rice with beans* (pattern iii) is stronger than *eat with beans* (pattern iv).

A3. Selectional preference evaluation of synonyms. Synonyms are generated for each word (except prepositions) in the pattern. Then, different combinations are tested against a previously acquired selectional preferences database—details on how to extract this database are described in Sect. 6.2.4. This database yields a *score* for a given combination. This score is calculated using a mutual information formula: $freq(w1, w2)/[freq(w1) + freq(w2) + freq(w1, w2)]$. Different formulae for calculating mutual information are presented in [122]. If the score of a combination is greater than a threshold, the combination is listed as a possible substitution. Some patterns may have more than one possible substitution, in which case each is listed in a particular order, e.g., starting from the highest selectional preferences and ending with the value closest to a given threshold. Additionally, the original construction is ranked using the same selectional preferences database.

A4. Enciphering. Each bit of information to be encoded decides which synonym paraphrasing will be done. Some patterns have several options for substitution, where each paraphrase may represent more than one bit. For example, given four possible substitutions, it is possible to represent four combinations of two bits— namely, 00, 01, 10, and 11.

A5. Re-agreement. If there are any substitutions that require simple syntactic structure changes, these are done now. For example, in Spanish, *historia* ("story") can be substituted by *cuento* ("tale"), but *historia* is feminine and *cuento* is masculine. Thus, it is necessary to change the article *la* (feminine "the") to *el* (masculine "the"), which results in *el cuento* and avoids *la cuento*.

At the receiver side, it is necessary to decode the hidden information, which is the task of a specific decoder-steganalyzer. The decoder-steganalyzer possesses the same resources as the encoder: the selectional preferences database and the tagging module. The text is tagged as in A1; the patterns are extracted as in A2. The synonym paraphrases are ranked as in A3; bits are extracted by mapping each possible combination in the same way as in A3 and A4. Re-agreement does not represent a problem during the decoding process, because articles and other words were discarded in A1.

6.2.6 A Manually Traced Example in Spanish

To illustrate the above-mentioned algorithm, we apply our method to hide a small amount of information in a fragment of Spanish text that has been extracted from a local newspaper[2]—see Fig. 6.5. The translation of this fragment is: "Sheltered in the Madison Square Garden to protect themselves against "terrorist" and protester's threats, Republicans began their celebration with self-congratulations of how they

---

[2]La Jornada, Mexico, August 2004.

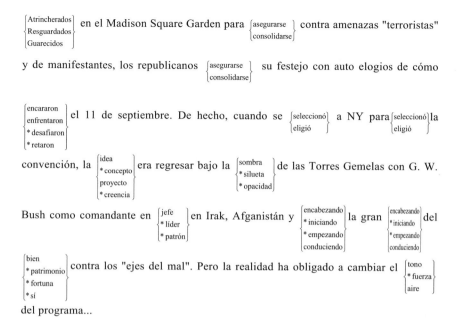

Fig. 6.5 Text with synonyms for paraphrasing—bad substitutions are marked with *

handled September 11. Indeed, when NY was selected to host the convention, the idea was to return under the shadow of the Twin Towers with G. W. Bush, as Commander in Chief of Iraq and Afghanistan, leading the great struggle of good against the "axes of evil." But reality necessitated that the program's angle change…".

We have listed several possible synonyms (obtained from a dictionary [113]) for various words in this example. Not every substitution is verifiable since our selectional preferences database does not contain every possible instance. Such is the case for combinations such as *Madison Square Garden*, for example. Other combinations—*asegurar(se) contra amenazas* ("ensure (themselves) against threats") versus *consolidar(se) contra amenazas* ("consolidate (themselves) against threats")—can be verified in our selectional preferences database: the first yields a score of 3 and the second yields a score of 0.2. If we set our threshold at around 0.5, the second option will be discarded.

Table 6.1 gives additional combinations of nouns that have been verified by the selectional preferences database. Entries not contained in the selectional preferences database are marked with a question mark ("?").

In Table 6.1, combinations above the threshold (0.5) are shown in bold. Alternative possibilities that allow the representation of one bit are marked in light gray; those that can represent two bits are marked with dark gray.

The fragment of text used for our example can hide eight bits (i.e., 1 byte) of information. The text has around 500 bytes. Thus, the ratio that measures

Table 6.1 Verified combinations and their score(s) from Fig. 3.2

word combination	s	word combination	s	word combination	s
atrincherar en Madison	?	creencia ser	0.31	conducir torneo	0.03
resguardar en Madison	?	sombra de torre	?	encabezar riña	0
guarecer en Madison	?	silueta de torre	?	empezar riña	0
asegurar contra amenaza	**3**	opacidad de torre	?	conducir riña	0
consolidar contra amenaza	0.2	**comandante en jefe**	**6.7**	lucha de bien	0.7
iniciar festejo	**0.7**	comandante en líder	0.45	lucha de patrimonio	0
comenzar festejo	**0.8**	comandante en patrón	0.4	lucha de fortuna	0
emprender festejo	0.4	jefe en Irak	?	lucha de sí	0
originar festejo	0	líder en Irak	?	torneo de bien	0
encarar 11	?	patrón en Irak	?	torneo de fortuna	0
enfrentar 11	?	**encabezar lucha**	**2.1**	torneo de sí	0
desafiar 11	?	**iniciar lucha**	**1.75**	riña de bien	0
retar 11	?	**conducir lucha**	**0.8**	riña de patrimonio	0
seleccionar a NY	**1.3**	**encabezar rivalidad**	**0.67**	riña de fortuna	0
elegir a NY	**1.2**	iniciar torneo	0.47	riña de sí	0
celebrar convención	**1.9**	empezar lucha	0.4	rivalidad de bien	0
hacer convención	**1.8**	empezar torneo	0.3	rivalidad de patrimonio	0
realizar convención	**1.6**	encabezar torneo	0.28	rivalidad de fortuna	0
festejar convención	**0.5**	iniciar rivalidad	0.08	rivalidad de sí	0
idea ser	**0.7**	iniciar riña	0.05	**aire de programa**	**0.66**
proyecto ser	**0.6**	empezar rivalidad	0.05	**tono de programa**	**0.54**
concepto ser	0.4	conducir rivalidad	0.05	fuerza de programa	0.23

steganographic bandwidth is approximately 0.002, which means that the text must be 500 times longer than the hidden information.

6.2.7 Conclusions

As with its previous version, the proposed method of linguistic steganography conserves the meaning of the carrier text as well as its inconspicuousness. One of the main advantages of our method is that it does not require a manually collected database of collocations. Instead, automatically extracted selectional preferences are used to acquire a large database. Since the method presented in this section is based on automatically acquired resources, it is possible to extend its application to many languages, e.g., Spanish. The results are less refined than those obtained with manually collected collocations, but they seem acceptable.

On the other hand, the mean steganographic bandwidth value of 0.002, which is obtained with local synonymous paraphrasing, seems rather low. An example of the maximum synonym paraphrasing that can be reached is provided in [15]. That study argues that, by starting from the samples of synonymous paraphrasing given by I. Mel'čuk, the maximum steganographic bandwidth can reach approximately 0.016; this value can be obtained by considering synonym paraphrasing for complete phrases, such as *ayudar* ("to help") and *dar ayuda* ("to give help"). Compiling a list of such phrases that can be substituted for one another is currently in the very early stages of development. Our method reaches 12.5% of the maximum possible level without considering adjective variants. Thus, the obtainable value of synonym paraphrasing bandwidth evidently depends on how saturated the linguistic resources are. For this reason, such resources should be developed further, without any clear-cut limits of perfection. In particular, the bandwidth value obtained by our method can be improved by taking adjective variants into consideration; this is, thus, part of our planned future work.

As to our algorithm, we hardly consider it faultless. The following issues now seem especially acute:

- Large chains of word combinations, such as *encabezando la lucha del bien contra los "ejes del mal"* ("leading the great struggle of good against the 'axes of evil'") can lead to the wrong selection of synonyms since each combination is considered only by pairs, ignoring every other combination that makes up the whole.
- A large database of named entities is necessary in order to recognize phrases such as *el 11 de septiembre* ("September 11") or *Madison Square Garden*. Particularly, using the selectional preferences model can help because knowing that *Madison Square Garden* is a place helps to evaluate combinations such as *Atrincherados/resguardados/guarecidos en el Madison Square Garden* ("Entrenched/protected/sheltered/in Madison Square Garden").
- Threshold adjustments should be made automatically.

Each of these problems will be investigated in depth in future work.

Chapter 7
Prepositional Phrase Attachment Disambiguation

7.1 Using the Internet

The problem of disambiguating PP attachments consists of determining if a PP is part of a noun phrase (as in *He sees the room with books*) or a verb phrase (as in *He fills the room with books*). Volk has proposed two variants of a method that queries an Internet search engine to find the most probable PP attachment [194, 195]. Here, we apply the latest variant of Volk's method to Spanish, with several differences allowing us to attain better performance nearer that of statistical methods using treebanks.

7.1.1 Introduction

In many languages, PP, such as *in the garden*, can be attached to either noun phrases (NPs)—*the grasshopper in the garden*—or verb phrases (VPs)—*plays in the garden*. Sometimes several possibilities exist for a single PP attachment. For example, in *The police accused the man of robbery*, there are two possibilities: (1) the object is *the man of robbery*, or (2) the object is *the man* and the accusation is *of robbery*. English speakers know that the second option is the correct one, whereas a machine needs a method to automatically determine the correct option.

Several methods based on treebank statistics exist for finding the correct PP attachment place. These methods have been reported to achieve accuracy rates of up to 84.5% (see [28, 57, 77, 134, 162, 210]). However, resources like treebanks are not available for many languages and arc difficult to port; thus, a method that is less resource-demanding is desirable. Ratnaparkhi shows in [163] a method that requires only a PoS tagger and morphological information; in that method, raw text is used to train the tagger.

© Springer International Publishing AG 2018
A. Gelbukh and H. Calvo, *Automatic Syntactic Analysis Based on Selectional Preferences*, Studies in Computational Intelligence 765,
https://doi.org/10.1007/978-3-319-74054-6_7

The quality of the training corpus significantly determines the results' correctness. In particular, in order to reduce the effects of noise in a corpus and to consider most phenomena, a very large corpus is desirable. Eric Brill argues in [27] that it is possible to achieve state-of-the-art accuracy with relatively simple methods whose power comes from the plethora of text available to these systems. His study also gives examples of several natural language processing (NLP) applications that benefit from the use of very large copora.

Nowadays, large corpora comprise more than 100 million words, and the Web can be seen as the largest corpus with more than one billion documents. Particularly for Spanish, Bolshakov and Galicia-Haro report that approximately 12,400,000 pages can be found through Google [16]. We can consider the Web as a corpus big enough and diverse enough to obtain better results using statistical NLP methods.

Using the Web as corpus is a recently growing trend; a count of existing research that tries to harness the potential of the Web for NLP can be found in [108]. In particular, for the problem of finding the correct PP attachment, Volk [194, 195] proposes variants of a method that queries an Internet search engine to find the most probable PP attachment.

Here, we show the results of applying the latest variant of Volk's method with a few key differences to Spanish. In Sect. 7.1.2, we explain the variants of Volk's method. In Sect. 7.1.3, we discuss the differences between his method and ours. In Sect. 7.1.4, we provide details of our experiment and the results obtained. Finally, in Sect. 7.1.5, we present our conclusions.

7.1.2   Volk's Method

Volk proposes two variants of the same method to decide whether a PP attaches to an NP or VP. In this section, we describe both variants and discuss their results.

7.1.2.1   First Variant

In [194], Volk proposes uses the Web as a corpus to disambiguate PP attachments by considering the co-occurrence frequencies (freq) of verb+preposition constructions against those of noun+preposition constructions. The formula used to calculate these co-occurrence frequencies is:

```
cooc(X,P) = freq(X,P) / freq (X)
```

where X can be either a noun or a verb. For example, for *He fills the room with books*, N = *room*, P = *with*, and V = *fill*. `cooc(X,P)` is a value between 0 (no co-occurrences found) and 1 (co-occurrence always happens).

`freq (X,P)` is calculated by querying the Altavista search engine using the NEAR operator: `freq(X,P) = query("X NEAR P")`.

Table 7.1 Coverage and accuracy for Volk's 2000 algorithm

Threshold	Coverage (%)	Accuracy (%)
0.1	99.0	68
0.3	36.7	75
0.5	7.7	82

To choose the correct attachment, `cooc(N+P)` and `cooc(V+P)` are calculated and the higher value determines the attachment. If a `cooc` value is lower than a *minimum co-occurrence threshold*, the attachment cannot be decided; thus, it is not covered. By adjusting the *minimum co-occurrence threshold*, Volk's 2000 algorithm can attain either very good coverage and poor accuracy or very good accuracy and low coverage. Table 7.1 shows the coverage/accuracy values for Volk's experiments.

Volk also concludes, in [194], that using full forms is better than using lemmas.

Vandeghinste conducted the same experiment in Dutch [192] and obtained a coverage of 100% and an accuracy of 58.4%. To obtain an accuracy of 75%, Vandeghinste used a threshold of 0.606, which yielded a coverage of only 21.6%.

7.1.2.2 Second Variant

In a subsequent article [195], Volk used a different formula to calculate co-occurrences. Now, the head noun of the PP is included within the queries. The formula is as follows:

$$cooc(X,P,N_2) = freq(X, P, N_2) / freq(X)$$

`freq(X,P,N₂)` is calculated by querying the Altavista search engine using the `NEAR` operator: `freq(X,P,N₂) = query("X NEAR P NEAR N₂")`. X can be N_1 or V. For example, for *He fills the room with books*, N_1 = *room*, P = *with*, N_2 = *books* and V = *fill*.

Volk experiments first with requiring that both `cooc(N₁,P,N₂)` and `cooc(V,P,N₂)` be calculated to determine a result. Then, he considers using a threshold to determine the PP attachment if either `cooc(N₁,P,N₂)` or `cooc(V,P,N₂)` is unknown. That is, if `cooc(N₁,P,N₂)` is unknown, `cooc(V,P,N₂)` must be higher than the threshold to decide that the PP is attached to the verb and vice versa. Afterward, by including both lemmas and full forms in queries, Volk attains a better performance; furthermore, by defaulting to noun attachments for previously uncovered attachments, he attains a coverage of 100%. His results are shown in Table 7.2.

In Vandeghinste's Dutch experiments, requiring both `cooc(N₁,P,N₂)` and `cooc(V,P,N₂)` achieves a coverage of 50.2% with an accuracy of 68.92%. By using a threshold and including both lemmas and full forms in queries, he reaches 27% coverage with an accuracy of 75%. To obtain 100% coverage, Vandeghinste defaults previously uncovered cases to noun attachments, obtaining an accuracy of 73.08%.

Table 7.2 Results of Volk's 2001 method

Coverage (%)	Accuracy (%)	Requiring both cooc (N1,P, N2) and cooc(V,P,N$_2$)	Threshold when one of cooc (N1,P,N2) **or** cooc(V,P,N2) is not known	Includes both lemmas and full forms in queries	Defaults to noun attachment for uncovered attachments
55	74.32	✓	NA		
63	75.04		0.001		
71	75.59		0.001	✓	
85	74.23		0	✓	
100	73.08		0	✓	✓

Table 7.3 Number of co-occurrences found using different search engines

	leer en el metro	read in the subway
Google	104	30
All-the-web	56	23
Altavista	34	16
Teoma	15	19

7.1.3   Improving Performance

Methods for resolving PP attachment ambiguity based on treebank statistics achieve, by far, better performance than do the experiments described above. Nonetheless, we believe that several elements could be changed to improve methods based on Web queries.

One such element is the size of the document databases of search engines—indeed, this is relevant to obtaining representative co-occurrence frequencies for any given language. We know that not every search engine yields the same results. For example, Table 7.3 shows the number of co-occurrences found using different search engines for the same words.

A recent "search engine showdown"[1] lists Google as having the largest document database. Thus, we have determined that using Google to obtain word co-occurrence frequencies will likely yield better results.

Another element to consider is the use of the NEAR operator. We decided not to use it for our experiments because it does not guarantee that query words appear in the same sentence. Consider the following queries from Altavista:

wash NEAR with NEAR door	6,395 results	(1)
wash NEAR with NEAR bleach	6,252 results	(2)

(1) yields 6,395 pages found, even when books are unrelated to the wash operation, and (2) yields 6,252 pages found. Thus, we can see that there is not a clear

---

[1]www.searchengineshowdown.com

distinction for when a preposition+noun is related to a verb. On the other hand, using an exact phrase search yields 0 results, marking a clear distinction between "wash with door" and "wash with bleach." The results found are as follows:

Exact phrase search	Results	Search engine
"wash with door"	0	Altavista
"wash with bleach"	100	Altavista
"wash with door"	0	Google
"wash with bleach"	202	Google

Following [195], we use jointly full forms and lemmatized forms of nouns and verbs to obtain better performance. However, since we are not using the NEAR operator, we must consider the determiners that can be placed between the noun or verb and the preposition. Also, we consider that the nucleus of the PP might appear in plural form without affecting its use. To illustrate this, consider the following sentence[2]:

Veo al gato con un telescopio ("I see the cat with a telescope")

The attachments are calculated by the queries shown in Table 7.4.

Since freq(veo,con,telescopio) is higher than freq(gato,con, telescopio), the attachment is given to *veo con telescopio*.

7.1.4 Experiment and Results

For our evaluation, we randomly extracted 100 sentences from the LEXESP corpus of Spanish [64] and the newspaper Milenio Diario.[3] All searches were restricted to only pages in Spanish.

First, we considered not restricting queries to a specific language, given that a benefit could be obtained from retrieving similar words across languages—such as French and Spanish. For example, the phrase *responsables de la debacle* ("culprits of the collapse") is used in both languages with the only difference being the word accentuation for *debacle* (*débâcle* in French, *debacle* in Spanish). As Google does not take into account word accentuation, results for both languages are returned by the same query. However, with an unrestricted search, Google returns different counts in its API[4] than it does in its GUI.[5] For example, for *ver* ("to see"), Google's GUI returns 270,000 results, whereas its API returns more than 20,000,000, even with the "group similar results" filter turned on. This enormous deviation can be reduced by restricting queries to a specific language. By restricting the query to

---

[2]Example borrowed from [79].

[3]www.milenio.com.

[4]Google API is a web service that uses the SOAP and WSDL standards to allow a program to directly query the Google search engine. More information can be found at api.google.com .

[5]www.google.com.

Spanish, a search for *ver* ("to see") returns 258,000 results in Google's GUI, whereas the API returns 296,000. We do not know the reason for this difference, but it does not have a significant impact on our experiment (Table 7.5).

Table 7.4 Queries to determine the PP attachment of *Veo al gato con un telescopio* and *I see the cat with a telescope*

Veo al gato con un telescopio	Hits	I see the cat with a telescope	Hits
ver	296,000	see	194,000,000
"ver con telescopio"	8	"see with telescope"	13
"ver con telescopios"	32	"see with telescopes"	76
"ver con un telescopio"	49	"see with a telescope"	403
"ver con el telescopio"	23	"see with the telescope"	148
"ver con unos telescopios"	0	"see with some telescopes"	0
"ver con los telescopios"	7	"see with the telescopes"	14
veo	642,000		
"veo con telescopio"	0		
"veo con telescopios"	0		
"veo con un telescopio"	0		
"veo con unos telescopios"	0		
"veo con el telescopio"	1		
"veo con los telescopios"	0		
freq(veo,con,telescopio) =	**1.279×10^{-4}**	**freq(see,with,telescope) =**	**3.371×10^{-6}**
gato	185,000	cat	24,100,000
"gato con telescopio"	0	"cat with telescope"	0
"gato con telescopios"	0	"cat with telescopes"	0
"gato con un telescopio"	3	"cat with a telescope"	9
"gato con unos telescopios"	0	"cat with some telescopes"	0
"gato con el telescopio"	6	"cat with the telescope"	2
"gato con los telescopios"	0	"cat with the telescopes"	0
freq(gato,con,telescopio) =	**0.486×10^{-4}**	**freq(cat,with,telescope) =**	**0.456×10^{-6}**

Table 7.5 Occurrence examples for some verbs in Spanish

Triple	Literal English translation	Occurrences	% of total verb occurrences (%)
ir a {actividad}	go to {activity}	711	2.41
ir a {tiempo}	go to {time}	112	0.38
ir hasta {comida}	go until {food}	1	0.00
beber {sustancia}	drink {substance}	242	8.12
beber de {sustancia}	drink of {substance}	106	3.56
beber con {comida}	drink with {food}	1	0.03
amar a {agente_causal}	love to {causal_agent}	70	2.77
amar a {lugar}	love to {place}	12	0.47
amar a {sustancia}	love to {substance}	2	0.08

The sentences used in our experiment contain 181 cases of preposition attachment ambiguity. From those, 162 could be automatically resolved; these were verified manually, and it was determined that 149 were resolved correctly and 13 incorrectly.

We obtained a coverage of 89.5% with an accuracy of 91.97%, compared with the coverage and accuracy obtained by Volk. Without considering coverage, the overall percentage of attachment ambiguities resolved correctly is 82.3%.

7.1.5 Conclusions

We have managed to improve performance using Volk's method by implementing the following differences: using exact phrase searches instead of the NEAR operator, using a search engine with a larger document database, searching combinations of words that include definite and indefinite articles, and searching for both the singular and plural forms of words when possible. The results obtained with this method (89.5% coverage, 91.97% accuracy, 82.3% overall) are very close to those obtained using treebank statistics, without the need of such resources. Our method can be tested at likufanele.com/ppattach.

7.2 PP Attachment Disambiguation Using Selectional Preferences

Extracting information automatically from texts for database representation requires previously well-grouped phrases so that entities can be separated adequately. This problem is known as PP attachment disambiguation. Current PP attachment disambiguation systems either require an annotated treebank or use an Internet search engine to achieve a precision of more than 90%. Unfortunately, such resources are not always available. In addition, using the same techniques that use the Web as a corpus may not achieve the same results when using local corpora. Here, we present an unsupervised method for generalizing local corpora information by means of the semantic classification of nouns based on the top 25 unique beginner concepts of WordNet. We then propose a method that uses this information for PP attachment disambiguation.

Extracting information automatically from texts for a structured representation requires previously well-grouped phrases so that entities can be separated adequately. For example, in the sentence *See the cat with a telescope*, two different groupings are possible: *See [the cat] [with a telescope]* or *See [the cat with a telescope]*. The first case involves two different entities, while the second case has a single entity. This problem is known in syntactic analysis as PP attachment disambiguation.

There are several methods for disambiguating a PP attachment. Earlier methods, e.g., those described in [28, 162], achieve accuracy rates of up to 84.5% using treebank statistics. Kudo and Matsumoto [113] obtained 95.77% accuracy with an algorithm that required weeks of training, and Lüdtke and Sato [120] achieved 94.9% accuracy with one that only required three hours. Both methods require a corpus that is annotated syntactically with chunk marks; however, this kind of corpus is not available for every language and the cost of building one is rather high, considering the number of person-hours that are required. A method that works with untagged text is presented in [36]. That method has an accuracy rate of 82.3% and uses the Web as a corpus, which means that it can be slow—up to 18 queries are used to resolve a single PP attachment ambiguity, and each preposition +noun pair found in a sentence multiplies that number.

The algorithm presented in [36] is based on the idea that a very big corpus has enough representative terms to allow PP attachment disambiguation. Since it is possible to have very big corpora locally these days, we ran experiments to explore the possibility of applying such a method without an Internet connection. We tested with a very big corpus of 161 million words in 61 million sentences. This corpus was obtained online and pulled three years' of publication from four newspapers. The results were disappointing—the same algorithm that used the Web as a corpus to yield a recall of almost 90% had a recall of only 36% and a precision of almost 67% using this local newspaper corpus instead of the Web.

Therefore, we hypothesize that the information contained in the local newspaper corpus needs to be generalized in order to maximize recall and precision. To do so, selective preferences—which measure the probability that a complement is used for a certain verb, based on the semantic classification of the complement—may be used. In this way, the problem of analyzing *I see the cat with a telescope* can be solved by instead considering *I see {animal} with {instrument}*.

For example, to disambiguate the PP attachment for the sentence *Bebe de la jarra de la cocina* ("(he) drinks from the jar of the kitchen"), selectional preferences provide information such as the knowledge that *from {place}* is an uncommon complement for the verb *bebe* ("drinks") and, thus, the probability of attaching this complement to *bebe* is low. Therefore, it is attached instead to the noun *jarra*, yielding *Bebe de [la jarra de la cocina]* ("(he) drinks [from the jar of the kitchen]") (Fig. 7.1).

Table 7.8 gives additional occurrence examples for some Spanish verbs. From this table, we can see that the verb *ir* ("to go") is mainly used with the complement *a {activity}* ("to {activity}"). Lesser used combinations have almost zero occurrences, such as *ir hasta {food}* (which literally means "go until {food}"). Lastly, the verb *amar* ("to love") is often used with the preposition *a* ("to").

In this chapter, we propose a method to obtain selectional preferences information such as that shown in Table 7.6. In Sect. 7.2.1, we will briefly discuss related work on selectional preferences. Sections 7.2.2–7.2.4 explain our method. In Sect. 7.2.5, we present an experiment and evaluation of our method as applied to PP attachment disambiguation. Finally, we conclude the chapter. Our conclusions and plans for future work in this area are discussed in Sect. 7.2.6.

{food}:   breakfast, feast, cereal, beans, milk, etc.
{activity}:   abuse, education, lecture, fishing, hurry, test
{time}:   dawn, history, Thursday, middle age, childhood
{substance}:   alcohol, coal, chocolate, milk, morphine
{name}:   John, Peter, America, China
{causal_agent}:   lawyer, captain, director, intermediary, grandson
{place}:   airport, forest, pit, valley, courtyard, ranch

Fig. 7.1 Examples of words for the categories shown in Table 7.1

Table 7.6 Examples of semantic classifications of nouns

Word	English translation	Classification
rapaz	predatory	activity
rapidez	quickness	activity
rapiña	prey	shape
rancho	ranch	place
raqueta	racket	thing
raquitismo	rickets	activity
rascacielos	skyscraper	activity
rasgo	feature	shape
rastreo	tracking	activity
rastro	track	activity
rata	rat	animal
ratero	robber	causal agent
rato	moment	place
ratón	mouse	animal
raya	boundary	activity
raya	manta ray	animal
raya	dash	shape
rayo	ray	activity
raza	race	grouping
razón	reason	attribute
raíz	root	part
reacción	reaction	activity
reactor	reactor	thing
real	real	grouping
realidad	reality	attribute
realismo	realism	shape
realización	realization	activity
realizador	producer	causal agent

7.2.1 Related Work

The terms *selectional constraints* and *selectional preferences* are relatively new, although similar concepts are present in works such as [71, 199]. One of the earliest works using these terms is [166], where Resnik considered selectional constraints to determine the restrictions that a verb imposes on its object. Selectional constraints have rough values, such as whether an object of a certain type can be used with a verb. Selectional preferences are graded and measure, for example, the probability that an object is used for a given verb [167]. Such works use a shallow parsed corpus and a semantic class lexicon to find selectional preferences for word sense disambiguation.

Another work that uses semantic classes for syntactic disambiguation is [160]. In that work, Prescher et al. use an EM clustering algorithm to obtain a probabilistic lexicon based on classes. This lexicon is used to disambiguate target words in automatic translation.

A work that particularly uses WordNet classes to resolve PP attachment is [28]. In that work, Brill and Resnik apply the transformation-based error-driven learning model to disambiguate the PP attachment, obtaining an accuracy of 81.8%; theirs is a supervised algorithm.

As far as we know, selectional preferences have not yet been used in unsupervised models for PP attachment disambiguation.

7.2.2   Sources of Noun Semantic Classification

A semantic classification for nouns can be obtained from existing WordNets by using a reduced set of classes that correspond to the unique beginners for WordNet nouns described in [90]. These classes are activity, animal, life_form, phenomenon, thing, causal_agent, place, flora, cognition, process, event, feeling, form, food, state, grouping, substance, attribute, time, part, possession, and motivation. To these unique beginners, name and quantity are added. The name class corresponds to capitalized words that are not found in the semantic dictionary and quantity class corresponds to numbers.

Since not every word is covered by WordNet and since there is not a WordNet for every language, the semantic classes can be alternatively obtained automatically from human-oriented explanatory dictionaries; a method for doing so is explained in detail in [37]. Examples of semantic classification of nouns extracted from a human-oriented explanatory dictionary [115] using this method are shown in Table 7.8.

7.2.3   Preparing Sources for Selectional Preferences
Extraction

Journals and newspapers are common sources for large amounts of medium-to-good quality text. However, these media tend to express many ideas in a small amount of space.

This tendency leads to long sentences that are full of subordinate clauses, especially for languages that allow an unlimited number of such clauses to be nested together. Therefore, one of the first problems to be solved is how to break a sentence into several sub-sentences. Consider, for example, the sentence shown in Fig. 7.2—it is a single sentence extracted from a Spanish newspaper.

We use two kinds of delimiters to separate subordinate sentences: delimiter words and delimiter patterns. Examples of delimiter words are *pues* ("well"), *ya que* ("given that"), *porque* ("because"), *cuando* ("when"), *como* ("as"), *si* ("if"), *por eso* ("because of that"), *y luego* ("and then"), *con lo cual* ("with which"), *mientras* ("in the meantime"), *con la cual* ("with which") (feminine), and *mientras que* ("while").

Y ahora, cuando
     (el mundo) **está** gobernado por (las leyes del mercado),
cuando
     (lo determinante en la vida) **es**
     comprar o
     vender, sin
     fijarse en <los que
     carecen de todo>,
     son fácilmente **comprensibles** <las razones de
       <la ola de publicidad global que
       convenció <a los posibles compradores de servicios y
regalos > de que
         había (grandes razones) para
         celebrar> y
     como les **pareciese** poco (el fin de año)
     se **lanzaron** a
     propagar (el fin del siglo y del milenio)

Literal English translation:

And now, when
     the world is governed by market's laws, when
     what **determines** life **is**
     to buy or
     to sell without
     taking into account those that
     don't **have** anything,
     easily understandable **are** the reasons for
       the global publicity wave that
       convinced the possible buyers of services and gifts that
       there were great reasons to
       celebrate, and
     as the end of the year **was** not enough for them,
     they **launched** themselves
     to propagate the end of the century and the millennium

Fig. 7.2 Very long sentence in a style typically found in journals. () surround simple NPs; < > surround NP subordinate clauses, verbs are in boldface

PREP V ,	CONJ PRON V	CONJ N V
V ADV que	PREP DET que N	PREP DET V
, PRON V	N que V	, N V
V PREP N , N V	, donde	N , que V
V PREP N , N PRON V	N , N	N , CONJ que
V PREP N V	CONJ N N V	N que N PRON V
V de que	CONJ N PRON V	CONJ PRON que V V

Fig. 7.3 Delimiter patterns. V verb, PREP preposition, CONJ conjunction, DET determiner, N noun, and lowercase indicates strings of words

Examples of delimiter patterns are shown in Fig. 7.3; those patterns are PoS-based, so the text was shallow parsed prior to their application.

The sentence in Fig. 7.2 was separated using this simple technique so that each sub-sentence would lie in a different row.

7.2.4 Extracting Selectional Preferences Information

Now that sentences are tagged and separated, our purpose is to find the following syntactic patterns:

1. Verb $_{NEAR}$ Preposition $_{NEXT\_TO}$ Noun
2. Verb $_{NEAR}$ Noun
3. Noun $_{NEAR}$ Verb
4. Noun $_{NEXT\_TO}$ Preposition $_{NEXT\_TO}$ Noun

Patterns 1–3 will be referred to henceforth as *verb patterns*. Pattern 4 will be referred as a *noun* or *noun classification pattern*. The $_{NEAR}$ operator indicates that there might be other words in between. The operator $_{NEXT\ TO}$ indicates that there are no words in between. Note that word order is preserved; thus, pattern 2 is different from pattern 3. The results of these patterns are stored in a database. For verbs, the lemma is stored. For nouns, its semantic classification, when available through Spanish WordNet, is stored. Since a noun may have several semantic classifications due to, for example, having several word senses, a different pattern is stored for each semantic classification. For example, see Table 7.8, which shows the information extracted for the sentence in Fig. 7.2 (Table 7.7).

Once this information is collected, the occurrence of patterns is counted. For example, the last two rows in Table 7.8 (*fin, de, año* and *fin, de, siglo*) add two of each of the following occurrences: place of cognition, cognition of cognition, event of cognition, time of cognition, place of time, cognition of time, event of time, and time of time. (An example of the type of information that results from this process is shown in Table 7.8.) This information is then used as a measur e of the selectional preference that a noun has to either a given verb or to another noun.

Table 7.7 Semantic patterns information extracted from the sentence given in Fig. 7.2

Words	Literal translation	Pattern
gobernado, por, ley	governed, by, law	*gobernar, por*, cognition
gobernado, de, mercado	governed, of, market	*gobernar, de*, activity thing
es, en, vida	is, in, life	*ser, en*, state life_form causal_agent attribute
convenció, a, comprador	convinced, to, buyer	*convencer, a*, causal_agent
convenció, de, servicio	convinced, of, service	*convencer, de*, activity process possession thing grouping
pareciese, de, año	may seem, of, year	*parecer, de*, cognition time
lanzaron, de, año	released, of, year	*lanzar, de*, cognition time
propagar, de, siglo	propagate, of, century	*propagar, de*, cognition time
propagar, de, milenio	propagate, of, millennium	*propagar, de*, cognition time
ley, de, mercado	law, of, market	cognition, *de*, activity thing
ola, de, publicidad	wave, of, publicity	event, *de*, activity cognition
comprador, de, servicio	buyer, of, service	causal_agent, *de*, activity process possession thing grouping
fin, de, año	end, of, year	place cognition event time, *de*, cognition time
fin, de, siglo	end, of, century	place cognition event time, *de*, cognition time

7.2.5   *Experimental Results*

The procedure explained in the previous sections was applied to a corpus of 161 million words comprising more than three years' worth of articles from four different Mexican newspapers. It took approximately three days on a Pentium IV PC to obtain 893,278 different selectional preferences for verb patterns (Patterns 1–3) for 5,387 verb roots and 55,469 different semantic selectional preferences for noun classification patterns (pattern 4).

7.2.5.1   PP Attachment Disambiguation

In order to evaluate the quality of the selectional preferences obtained, we tested their performance in PP attachment disambiguation. Consider the first two rows of Table 7.8, which correspond to the fragment *governed by the laws of the market*. That fragment reports two patterns of selectional preferences: *govern by {cognition}* and *govern of {activity/thing}*. With these obtained selectional preferences, it is possible to automatically determine the correct PP attachment: values of co-occurrence for *govern of {activity/thing}* and *{cognition} of {activity/thing}* are compared. The highest value sets the attachment.

Table 7.8 Results of PP attachment disambiguation using selectional preferences

File	#Sentences	Words	Average words per sentence	Kind of text	Precision (%)	Recall (%)
n1	252	4,384	17.40	News	80.76	75.94
t1	74	1,885	25.47	Narrative	73.01	71.12
d1	220	4,657	21.17	Sports	80.80	81.08
Total	546	10,926		Average	78.19	76.04

Formally, to decide if noun N_2 is attached to its preceding noun N_1 or is instead attached to verb V of the local sentence, the values of frequency for those attachments are compared using the following formula [195]:

$$freq(X, P, C_2) = \frac{occ(X, P, C_2)}{occ(X) + occ(C_2)}$$

where X can be V (a verb) or C_1 (the classification of the first noun N_1). P is a preposition and C_2 is the classification of the second noun N_2. If $freq(C_1, P, C_2) > freq(V, P, C_2)$, then the attachment is assigned to noun N_1. Otherwise, the attachment is assigned to verb V. The values of $occ(X, P, C_2)$ are the number of occurrences of the corresponding pattern in the corpus. See Table 7.6 for examples of verb occurrences. Examples of noun classification occurrences taken from the Spanish journal corpus are *{place} of {cognition}*: 354,213; *{place} with {food}*: 206; and *{place} without {flora}*: 21. The values of $occ(X)$ are the number of occurrences of the verb or noun classification in the corpus. For example, for *{place}*, the number of occurrences is 2,858,150.

7.2.5.2 Evaluation

The evaluation was conducted on three different files of the LEXESP corpus [179], which contains 10,926 words in 546 sentences. On an average, this method achieved a precision of 78.19% and a recall of 76.04%. Details for each file processed are shown in Table 7.9.

Table 7.9 State of the art for PP attachment disambiguation

Human (without context)		Use WordNet backoff		Use thesaurus backoff	
Ratnaparkhi [162]	88.2	Stetina and Nagao [186]	88.1	Pantel and Lin [151]	84.3
Mitchell [138]	78.3	Li and Abe 1998 [118]	85.2	McLauchlan [130]	85.0

7.2.6   Conclusions and Future Work

Using selectional preferences for PP attachment disambiguation yielded a precision of 78.19% and a recall of 76.04%. These results are not as good as the ones obtained with other methods, which had accuracies as high as 95%. However, this method does not require any expensive resources such as an annotated corpus or an Internet connection (to use the Web as a corpus); it does not even need to use a semantic hierarchy (such as WordNet) since the semantic classes can be obtained from human-oriented explanatory dictionaries, as discussed in Sect. 7.2.2.

We found also that, at least for this task, applying techniques that use the Web as a corpus to local corpora reduces the performance of those techniques more than 50% of the time, even if the local corpora are very big.

In order to improve results for PP attachment disambiguation using selectional preferences, our hypothesis is that instead of using only 25 fixed semantic classes, intermediate classes can be obtained using a whole hierarchy. In this way, it would be possible to have a flexible particularization for terms commonly used together, i.e., collocations such as *fin de año* ("end of year"), while maintaining the power of generalization. Another area for further development is the addition of a WSD module so that not every semantic classification for a single word is considered (see Sect. 6.1).

7.3   Applying Different Smoothing Methods

PP attachment can be addressed by considering the frequency counts of dependency triples contained within any given nonannotated corpus. However, even very large corpora fail to contain all possible triples. To solve this problem, several techniques have been used. Here, we evaluate two different backoff methods; one is based on WordNet and the other on a distributional (automatically created) thesaurus. Our evaluation uses Spanish. The thesaurus is created using the dependency triples found in the same corpus that is used to count the frequency of unambiguous triples. The training corpus used for both methods is an encyclopedia. The method based on a distributional thesaurus has higher coverage but lower precision than the method based on WordNet.

7.3.1   Theoretical Background

The PP attachment task can be illustrated by considering the canonical example *I see a cat with a telescope*. In this sentence, the PP *with a telescope* can be attached to either *see* or *cat*. Simple methods based on corpora address this problem by looking at the frequency counts of either word or dependency triples: *see with telescope* versus *cat with telescope*. In order to find enough occurrences of such triples, a very large corpus is needed. Such corpora are now available, and the Web may also be used [36, 195]; however, still some combinations of words do not occur. This is a familiar effect of Zipf's law: few words are very common and many words occur with low frequency [122]. The same is true for word combinations.

To address this problem, several backoff techniques have been explored. In general, "backing off" comprises looking at statistics for a set of words when there is insufficient data for a particular word. Thus, *cat with telescope* turns into ANIMAL *with* INSTRUMENT and *see with telescope* turns into *see with* INSTRUMENT (capitals denote sets of words). One way to identify the set of words associated with a given word is to use WordNet; another is to use a distributional thesaurus. A distributional thesaurus is a thesaurus that is generated automatically from a corpus by finding words that occur in similar contexts [96, 184, 198]. Both approaches have already been explored (for English) and been shown to yield results close to human disambiguation (see Table 7.10).

Experiments using different techniques have been carried out independently; to date, there are no evaluations comparing WordNet to distributional thesauruses as we do here, according to [108]. To make this comparison, we use a single corpus for both cases, and the same corpus is used to generate both the thesaurus and WordNet generalizations; that same corpus is also used to count the number of dependency triples.

Our evaluation uses Spanish. This is, to the best of our knowledge, the first work exploring backoff methods for PP attachments in a language other than English.

Table 7.10 Different formulae for calculating VScore and NScore

Description		VScore	NScore
S	The simplest one	$|v, p, n_2|$	$|n_1, p, n_2|$
S2	Considering doubles too	$|v, p, n_2| \times |v, p, *|$	$|n_1, p, n_2| \times |n_1, p, *|$
LL3	Log likelihood ratio	See Fig. 7.2	
Feat	Simplified Roth features 171 and 198	$\log(|*, p, *|/|*, *, *|) +$ $\log(|v, p, n_2|/|*, *, *|) +$ $\log(|v, p, *|/|v, *, *|) +$ $\log(|*, p, n_2|/|*, *, n_2|)$	$\log(|*, p, *|/|*, *, *|) +$ $\log(|n_1, p, n_2|/|*, *, *|) +$ $\log(|n_1, p, *|/|v, *, *|) +$ $\log(|*, p, n_2|/|*, *, n_2|)$

7.3.2 PP Attachment with No Backoff

7.3.2.1 Building Resources

The main resource is the count of dependency triples (DTC). To increase coverage, instead of considering strictly adjacent words, we consider dependency relationships between word types (lemmas). Only unambiguous dependency relationships are considered. For example, the sentences *I see with a telescope* and *A cat with three legs is walking* will provide the dependency triples *see, with, telescope* and *cat, with, legs*, respectively. However, the sentence *I see a cat with a telescope* will not provide any dependency triple since it is an ambiguous case.

We extract all dependency triples from our corpus in a batch process. We first tag the text morphologically and then group adjectives with nouns and adverbs with verbs. Then, we search for the patterns *verb preposition noun, noun preposition noun, noun verb,* and *verb noun.* Determiners, pronouns, and other words are ignored.

According to Lin [119], dependency triples consist of two words and the grammatical relationship, including prepositions, between those two words in the input sentence. To illustrate the type of dependency triples extracted, consider a microcorpus (μC) consisting of two sentences: *A lady sees with a telescope* and *The lady with a hat sees a cat.* The triples corresponding to this μC are shown in Fig. 7.4. We then denote the number of occurrences of a triple $\langle w, r, w' \rangle$ as $|w, r, w'|$. From μC, $|lady, SUBJ, see| = 2$ and $|lady, with, hat| = 1$. $|*, *, *|$ denotes the total number of triples (10 in μC), and an asterisk $*$ represents any word or relationship. In μC, $|see, *, *| = 4$, $|*, with, *| = 2$, and $|*, *, lady| = 2$ (Table 7.10).

Grammatical relationships without prepositions will be useful later for building the thesaurus, where word similarity will be calculated based on the contexts shared between two words. Now, however, we use DTC only to count *verb–preposition–noun$_2$* and *noun$_1$–preposition–noun$_2$* triples to determine a PP attachment, as explained in the following section.

7.3.2.2 Applying Resources

The task is to decide the correct attachment of p, n_2 given a 4-tuple of *verb noun$_1$ preposition noun$_2$*: (v, n_1, p, n_2). The attachment of p, n_2 can be either to the verb v or the noun n_1. The simplest unsupervised algorithm selects the attachment

see, SUBJ, lady	see, SUBJ, lady	see, OBJ, cat
lady, SUBJ-OF, see	lady, SUBJ-OF, see	cat, OBJ-OF, see
see, with, telescope	lady, with, hat	
telescope, with_r, see	hat, with_r, lady	

Fig. 7.4 Dependency triples extracted from μC

$$x = \left|x,*,*\right| \quad p = \left|*,p,*\right| \quad n = \left|*,*,n_2\right| \quad t = \left|*,*,*\right| \quad \bar{x} = t - x, \quad \bar{p} = t - p, \quad \bar{n} = t - n$$

$$xpn = \left|x,p,n_2\right|, \quad \bar{x}pn = \left|*,p,n_2\right| - xpn, \quad x\bar{p}n = \left|x,*,n_2\right| - xpn, \quad xp\bar{n} = \left|x,p,*\right| - xpn$$

$$\overline{xp}n = n - xpn - \bar{x}pn - x\bar{p}n, \quad \overline{x}p\bar{n} = p - xpn - \bar{x}pn - xp\bar{n}$$

$$x\overline{pn} = x - xpn - x\bar{p}n - xp\bar{n}, \quad \overline{xpn} = t - \left(xpn + \bar{x}pn + x\bar{p}n + xp\bar{n} + \overline{xp}n + \overline{x}p\bar{n} + x\overline{pn}\right)$$

$$\begin{aligned}
score = {} & xpn \cdot \log\left[xpn/(x \cdot p \cdot n/t^2)\right] + \bar{x}pn \cdot \log\left[\bar{x}pn/(\bar{x} \cdot p \cdot n/t^2)\right] + \\
& x\bar{p}n \cdot \log\left[x\bar{p}n/(x \cdot \bar{p} \cdot n/t^2)\right] + xp\bar{n} \cdot \log\left[xp\bar{n}/(x \cdot p \cdot \bar{n}/t^2)\right] + \\
& \overline{xp}n \cdot \log\left[\overline{xp}n/(\bar{x} \cdot \bar{p} \cdot n/t^2)\right] + \overline{x}p\bar{n} \cdot \log\left[\overline{x}p\bar{n}/(\bar{x} \cdot p \cdot \bar{n}/t^2)\right] + \\
& x\overline{pn} \cdot \log\left[x\overline{pn}/(x \cdot \bar{p} \cdot \bar{n}/t^2)\right] + \overline{xpn} \cdot \log\left[\overline{xpn}/(\bar{x} \cdot \bar{p} \cdot \bar{n}/t^2)\right]
\end{aligned}$$

for VScore, x is v, for NScore, x is n_1

Fig. 7.5 Formulae for calculating the three-point log-likelihood

according to highest of VScore = $\left|v, p, n_2\right|$ or NScore = $\left|n_1, p, n_2\right|$. When both values are equal, we say that the attachment is undeterminable using this method.

The corpus used to count dependency triples in this experiment was the whole Encarta Encyclopedia 2004 in Spanish [136], which has 18.59 M tokens, 117,928 word types in 73 MB of text, 747,239 sentences, and 39,685 definitions. The corpus was tagged using the TnT tagger that was trained with the manually tagged (morphologically) corpus CLiC-TALP[6] and lemmatized using the Spanish Anaya dictionary [116].

Once the corpus is morphologically tagged and lemmatized, the dependency triples are extracted. Encarta produced 7 M dependency triple tokens, of which 3 M were different triples, i.e., there were 3 M dependency triple types. 0.7 M tokens (0.43 M triple types) involved prepositions.

We used four different formulae to calculate the VScore and NScore; these are listed in Table 7.11. The first two formulae can be seen as the calculus of the probability of each triple, e.g., $p(v, p, n_2) = \left|v, p, n_2\right|/\left|*, *, *\right|$. Since both VScore and NScore are divided by the same number $\left|*, *, *\right|$, it can be omitted without any difference. For the log-likelihood[7] formulae, see Fig. 7.5.

Following the PP attachment evaluation method used by Ratnaparkhi et al. [162], the task is to determine the correct attachment given a 4-tuple (v, n_1, p, n_2). We extracted 1,137 4-tuples, along with their correct attachment (N or V), from the manually tagged corpus Cast-3LB[8] [143]. Samples of these 4-tuples are shown in Table 7.11.

---

[6]http://clic.fil.ub.es. The TnT tagger trained with the CLiC-TALP corpus has a performance of over 92% [142].

[7]Log-likelihood was calculated using the Ngram Statistics Package. See [8].

[8]Cast-3LB is part of the 3LB project, financed by the Science and Technology Ministry of Spain. 3LB.

Table 7.11 Samples of the 4-tuples (v, n_1, p, n_2) used for evaluation

4-tuples	English gloss
informar comunicado del Banco_Central N	inform communication of Central_Bank N
producir beneficio durante periodo V	produce benefit during period V
defender resultado de elección N	defend results of election N
recibir contenido por Internet V	receive contents by Internet V
planchar camisa de puño N	iron shirt of cuff N

Table 7.12 Comparison of the formulae to calculate VScore and NScore

Method	Coverage	Precision
Baseline	1.000	0.661
S	0.127	0.750
S2	0.127	**0.773**
LL3	0.127	0.736
Feat	0.127	0.717

The baseline can be defined in two ways: (1) assign all attachments to $noun_1$, which gives a precision of 0.736, or (2) exclude all 4-tuples with the preposition *de* ("of"). The latter is based on the fact that the preposition *de* attaches to a noun in 96.9% of all 1,137 4-tuples found,[9] resulting in a precision of 0.855, which is high for a baseline since the human agreement level is 0.883. Thus, to avoid such a highly biased baseline, we opted to exclude all 4-tuples with the preposition *de*—no other preposition presents such a high bias. Our evaluations were then done using only 419 of the 1,137 4-tuples extracted. The baseline in this case comprises assigning all attachments to the verb, which provides a precision rate of 66.1%. The human intertagger agreement for 4-tuples excluding the preposition *de* is 78.7%, which is substantially lower than the human agreement for all 4-tuples. Our results are shown in Table 7.12.

The highest precision is provided by formula S2; thus, from now on, we will use that formula to compare results between the different backoff methods.

7.3.3 WordNet Backoff

7.3.3.1 Building the Dictionary

In order to determine the correct PP attachment, we need a wider coverage of dependency relationships. To that end, we constructed a dictionary that uses

---

[9]This is valid also for English. For the training set provided by Ratnaparkhi, the preposition *of* attaches to a noun in 99.5% of the 20,801 4-tuples.

WordNet in order to obtain generalizations of dependency relationships. For example, *eat with fork*, *eat with spoon*, and *eat with knife* are generalized into *eat with {tableware}*. Note that *{tableware}* is not a word but rather a concept in WordNet, which provides the knowledge that *fork*, *spoon*, and *knife* are *{tableware}*. This way, if an unseen triple (such as *eat with chopsticks*) is found, WordNet can help by saying that *chopsticks* are *{tableware}* too, and thus, we can apply our knowledge about *eat with {tableware}* to *eat with chopsticks*.

Before we describe our method, let us introduce some notation. Every word w is linked to one or more synsets in WordNet, according to its different senses. W_n denotes the synset corresponding to the n-th sense of w, and N denotes the total number of senses. The hypernyms of each of these synsets provide several paths to the root. W_n^m denotes the m-th hypernym of the n-th sense of w', and M_n denotes the depth, i.e., the number of hypernyms to the root for sense number n.

For example, *glass* has seven senses in WordNet. The third hypernym of the fourth sense of *glass* is denoted by $W_4^3 = astronomical\_telescope$. An extract for *glass* from WordNet is given below for illustration:

Sense 2: *glass* (drinking glass) $\rightarrow$ container $\rightarrow$ instrumentality $\rightarrow$ artifact $\rightarrow$ object $\rightarrow$ whole $\rightarrow$ object $\rightarrow$ entity

Sense 4: *glass* (spyglass) $\rightarrow$ refracting_telescope $\rightarrow$ optical_telescope $\rightarrow$ **astronomical_telescope** $\rightarrow$ telescope $\rightarrow$ magnifier $\rightarrow$ scientific_instrument $\rightarrow$ instrument $\rightarrow$ device $\rightarrow$ instrumentality $\rightarrow$ artifact $\rightarrow$ object $\rightarrow$ entity

Our WordNet backoff method is based on [38, 54]. To extend a score (either NScore or VScore) through WordNet, we must consider all triples that involve the same w and r, varying w' (as in the case of learning *eat with {tableware}* from several examples of *eat with ***). This set of triples is denoted by $\langle w, r, * \rangle$. For each involved w', we evenly distribute[10] each score $s(w, r, w')$ among each of its senses of w' (as in [167]), and the result is then propagated to all hypernyms W_n^m. The resulting value is accumulative: higher nodes in WordNet collect information from all their daughters. In this way, general concepts summarize the usage (frequency of triples) of specific concepts (hyponyms).

To avoid overgeneralization (that is, the excessive accumulation at top levels,), the depth of the hypernyms must be considered. Sometimes the depth of the hypernyms' chain is very large (as in that of Sense 4 for *glass*), and sometimes it is small (as in that of Sense 2 for *glass*). A useful propagation formula that allows generalization and considers the depth of hypernym chains is given below (Table 7.13):

$$s(w, r, W_n^m) = [s(w, r, w')/N] \times [1 - (m - 1/M_n)] \tag{7.1}$$

---

[10]We assume an equiprobable distribution, which is problematic. However, there are currently no comprehensive sense-tagged texts for Spanish from which to extract sense distributions.

Table 7.13 Examples of relationship triples (w, r, w') with WordNet backoff

w	r	w'	English	Score
comer	con	mano	hand	3.49
'eat'	'with'	cubiertos	cutlery	1.31
		tenedor	fork	1.19
matar	con	arma	weapon	0.27
'kill'	'with'	armamento	armaments	0.23
		utillaje	utensil	0.18

In addition, the number of triples contributing to a certain WordNet node is counted for averaging at the upper levels. That is, after considering k triples $\langle w, r, * \rangle$, we count the number of triple types that contribute to each node. Then, the value of each node is divided by that number.

An illustration of our algorithm is shown in Fig. 7.6. Suppose we only have three triples, each of which is listed along with its count in Fig. 7.6. The frequency count for each triple is added to the corresponding word in WordNet. For *eat with fork*, the node for the word *fork* is labeled with three counts for *eat with. fork* may be used with other combinations of words, but we show here only those values associated with *eat with,* i.e., $\langle w, r, * \rangle$. According to Formula (7.1), the value obtained by labeling fork with three counts is divided by the number of senses that *fork* has. In this example, we assume two different senses of *fork*, each with its own hypernyms: {*division*} and {*cutlery*}. Focusing on the {*cutlery*} branch, we see how this value is propagated toward {*entity*}. For this branch, there are five levels of depth from {*entity*} to *fork* ($M_2 = 5$)—the other branch has four levels ($M_1 = 4$). Following the propagation of *fork* up the tree, we see how each level has a lower weight factor—that for {*tableware*} is 3/5 and that for {*entity*} is only 1/5. Each node is accumulative; therefore, {*cutlery*} accumulates the values for *fork, knife,* and *spoon*. The value for {*cutlery*} is divided by three because that is the number of types of contributing triples. If we had another triple, such as *eat with chopsticks,* then {*cutlery*} would remain untouched, but {*tableware*} would be divided by four.

For our experiment, we used Spanish EuroWordNet[11] 1.0.7 (S-EWN) [74], which has 93,627 synsets (62,545 nouns, 18,517 adjectives, 12,565 verbs), 51,593 hyponym/hypernym relationships, 10,692 meronym relationships, and 952 role information entries (noun agent, instrument, location, or patient). We propagated all dependency triples in DTC using Formula (7.1) (the creation of DTC is explained in Sect. 7.3.2.1).

The WordNet backoff algorithm presented in this section produces subjectively good results. Table 7.14 lists the top three qualifying triples in which *con* has a relationship with two common Spanish verbs.

---

[11]S-EWN was developed jointly by the University of Barcelona (UB), the National University of Open Education (UNED), and the Polytechnic University of Cataluña (UPC), Spain.

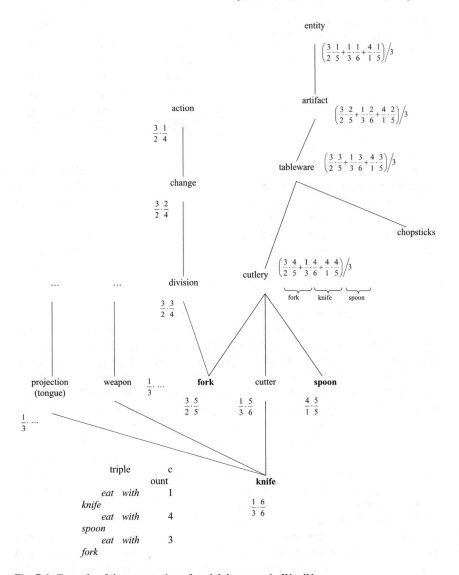

Fig. 7.6 Example of the propagation of a triple's counts in WordNet

7.3.3.2 Using the Dictionary

To decide a PP attachment in a 4-tuple (v, n_1, p, n_2), we calculate NScore for (n_1, p, n_2), and VScore for (v, p, n_2) as in Sect. 7.3.2.2. The highest score determines the attachment. WordNet backoff is applied when a triple is not found. In this case, n_2 is substituted by its hypernyms until the score from the new triple (x, p, W_n^m) is found in the previously calculated WordNet-extended scores. When calculating NScore, x is n_1, and when calculating VScore, x is v. The highest score determines the

Table 7.14 Example of similar words using Lin's similarity method

Word w	Similar word w'	English	$sim_{lin}(w, w')$
guitarrista	*pianista*	pianist	0.141
'guitarist'	*fisiólogo*	physiologist	0.139
	educador	teacher	0.129
devoción	*afecto*	affection	0.095
'devotion'	*respeto*	respect	0.091
	admiración	admiration	0.078
leer	*editar*	to edit	0.078
'to read'	*traducir*	to translate	0.076
	publicar	to publish	0.072

attachment. Note that we are backing off only n_2. We decided not to back off v because the verb structure in S-EWN has very few hypernym relations for verbs (7,172), and the definition of a hypernym for a verb is not clear in many cases. Since we do not back off v, we cannot back off n_1 as this would introduce a bias of NScores against VScores. Moreover, note that W_n^m is a specific synset in the WordNet hierarchy, and hence it has a specific sense. The problem of disambiguating the sense of n_2 is solved by choosing the highest value from each set of senses in each hypernym layer; see [38, 186] for WSD using PP attachment information. Results for this method will be presented in Sect. 5.

Following the example from Fig. 7.6, suppose we want to calculate the VScore for *eat with chopsticks*. Since this triple is not found in our corpus of frequency counts, we search for the hypernyms of *chopsticks*, in this case, {*tableware*}. Then, the value of this node is used to calculate VScore.

7.3.4 Thesaurus Backoff

7.3.4.1 Building the Dictionary

Here, we describe the automatic building of a thesaurus so that words not found in the dependency triples can be substituted by similar words. This similarity measure is based on Lin's work [151]. This thesaurus is based on the similarity measure described in [119]. The similarity between two words w_1 and w_2 as defined by Lin is as follows:

$$sim_{lin}(w_1 w_2) = \frac{\sum_{(r,w) \in T(w_1) \cap T(w_2)} (I(w_1, r, w) + I(w_2, r, w))}{\sum_{(r,w) \in T(w_1)} I(w_1, r, w) + \sum_{(r,w) \in T(w_2)} I(w_2, r, w)}$$

$$I(w_1, r, w') = \log \frac{|w, r, w| \times |*, r, *|}{|w, r, *| \times |*, r, w'|}$$

$T(w)$ is the set of pairs (r, w') such that $I(w, r, w')$ is positive. The algorithm for building the thesaurus is as follows:

Table 7.15 Experimental results for PP attachment disambiguation

Method	Coverage	Precision	Average
Manual agreement (human)	1.000	0.787	0.894
Default to verb (baseline)	1.000	0.661	0.831
No backoff	0.127	**0.773**	0.450
WordNet backoff	0.661	0.693	0.677
Distributional thesaurus backoff	**0.740**	0.677	**0.707**

```
for each word type w1 in the corpus
    for each word type w2 in the corpus
        sims(w1) ← {simlin(w1,w2), w2}
    sort sims(w1) by similarity in descending order
```

Like the WordNet method, this gives subjectively satisfactory results:
Table 7.15 lists the three most similar words to *guitarrista* "guitarrist," *devoción*
"devotion," and *leer* "to read."

7.3.4.2   Using the Dictionary

To determine a PP attachment in a 4-tuple (v, n_1, p, n_2), our algorithm calculates the
NScore for (n_1, p, n_2) and the VScore for (v, p, n_2), as in Sect. 7.3.2.2. The highest
score determines the attachment. When a triple is not found, the backoff algorithm is
applied. In this case, n_2 is substituted by its most similar word n'_2, which is calculated
using $sim_{lin}(n_2, n'_2)$. If the new triple (x, p, n'_2) is found in the count of dependency
triples (DTC), it is used to calculate the score. If it is not found, the next most similar
word is tried as a substitution; this continues until the new triple (x, p, n'_2) is found.
When calculating NScore, x is n_1; when calculating VScore, x is v. The highest score
determines the attachment. When n = 1, the n-th most similar word corresponds to
the first most similar word—for example, *pianist* for *guitarist*. For n = 2, the first
most similar word would be *physiologist* and so on. The algorithm is given below.

```
To decide the attachment in (v,n1,p,n2):
    VSCore = count(v,p,n2)
    NScore = count(n1,p,n2)
    n, m ← 1
    if NScore = 0
        while NScore = 0 & exists n-th word most similar to n2
            simn2  ← n-th word most similar to n2
            factor ← sim(n2,simn2)
            NScore ← count(n1,p,simn2) × factor
            n      ← n + 1
    if VScore = 0
        while VScore = 0 & exists n-th word most similar to n2
            simn2  ← m-th word most similar to n2
            factor ← sim(n2,simn2)
            VScore ← count(n1,p,simn2) × factor
            m      ← m + 1

    if NScore = VScore then cannot decide
    if NScore > Vscore then attachment is to n1
    if NScore < Vscore then attachment is to v
```

7.3.5 Comparison of Methods

In this section, we compare results of the three methods: no backoff, WordNet backoff, and thesaurus backoff. The results are listed in Table 10.4 along with baseline and manual agreement results. The third column shows the average between coverage and precision. Note that the baseline shown in Table 10.4 involves supervised knowledge: most attachments, after excluding *de* cases, are to the noun. The highest precision, coverage, and average values are in boldface. After excluding *de* cases, 419 cases remain. For 12.7%, all three algorithms have the same result; so the differences between WordNet backoff and distributional thesaurus backoff are based on the remaining 366 cases.

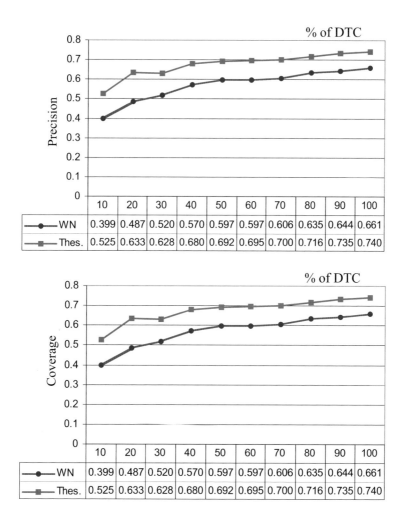

Fig. 7.7 Precision and coverage using different percentages of triple counts (0–100%)

Yet, not all cases are covered by these backoff methods—either because no substitution can be found for a certain word (as is the case for several acronyms or proper names) or because the triple cannot be found in DTC even after trying all possible substitutions. In general, the coverage we obtained is low because of the size of the corpus used to count attachment frequencies; although an encyclopedia provides many different words, the number of extracted prepositional attachments is rather low. We believe that using a bigger corpus will yield higher coverage measurements while maintaining the same relationship between the studied backoff methods, as suggested by our experiments, which only used randomly chosen partial percentages of the DTC corpus (see Fig. 7.7). Note that we used a totally unsupervised model. That is, we do not use any other backoff technique for uncovered cases in either algorithm.

7.3.6 Conclusions

Of the three methods evaluated for PP attachment, the best average measure was 0.707, which was obtained using the thesaurus backoff method, which has greater coverage than the other methods. However, that method has lower precision rates than does the WordNet backoff method. The no backoff method had very low coverage (0.127), but for the attachments that were covered, it had the best results: 0.773, which is close to manual agreement results. (Remember that this agreement is calculated after excluding the highly biased preposition *de* ("of"), which is practically always attached to nouns.) The WordNet backoff method's performance could be improved by adding information regarding the sense distribution of each word, rather than assuming an equiprobable distribution; however, doing so would push this method closer to being a supervised approach, and moreover, no resource providing such sense distributions is available for the Spanish language.

Our results indicate that an automatically built resource (in this case, a thesaurus) can be used instead of a manually built one and can still obtain similar results.

Our future work will explore using much larger corpora for counting triples and experiment with more sophisticated algorithms for determining attachments based on the thesaurus.

Chapter 8
The Unsupervised Approach: Grammar Induction

8.1 Introduction

There are mainly two approaches for creating syntactic dependency analyzers: supervised and unsupervised. The main goal of the first approach is to attain the best possible performance for a single language. For this purpose, a large collection of resources is gathered (using manually annotated corpora with part-of-speech annotations and syntactic and structure tags), which requires a significant amount of work and time. The state of the art in this approach attains syntactic annotation in about 85% of all full sentences [172]; in English, it attains over 90%. On the other hand, the unsupervised approach tries to discover the structure of a text using only raw text, which allows the creation of a dependency analyzer for virtually any language. Here, we explore this second approach. We present the model of an unsupervised dependency analyzer, named DILUCT-GI (GI short for grammar inference). We also propose adding morphological information both before and after the grammar induction process, thus converting shallow parsing to dependencies by reconstructing unavailable dependency information from the grammar inductors by means of a lexical categories precedence system, in a simpler fashion than that used in previous studies that implemented complex rule systems [53, 87, 169].

In the following sections, we present an overview of syntactic analyzers (Sects. 8.1.1–8.1.4), grammar induction algorithms (Sect. 8.2), and our system's implementation (Sects. 8.3 and 8.4). We also describe our method for converting constituent chunks into dependencies (Sect. 8.5), introduce the lexical categories precedence hierarchy (Sect. 8.5.1), and evaluate our method for use with both Spanish (Sect. 8.6) and English (Sect. 8.7). Our conclusions and discussion of future work are given in Sect. 8.8.

This chapter has been written with Omar Juárez-Gambino, ESCOM-IPN.

© Springer International Publishing AG 2018
A. Gelbukh and H. Calvo, *Automatic Syntactic Analysis Based on Selectional Preferences*, Studies in Computational Intelligence 765,
https://doi.org/10.1007/978-3-319-74054-6_8

8.1.1   Overview of Syntactic Analyzers

Recent syntactic analyzers have used manually tagged corpora to learn grammar structures [48]. Some analyzers have tried to learn grammar rules different from those used in the training corpus; however, these analyzers are usually less successful than the previously mentioned ones [30, 155].

Most of the analyzers in question use statistical techniques. Different probabilities are assigned to every possible representation of a sentence; the most probable representation is then selected and presented as the correct one. In the following sections, we present the state-of-the-art supervised, semisupervised, and unsupervised syntactic analyzers for comparison (Table 8.1).

8.1.2   Supervised Syntactic Analysis

Most state-of-the-art supervised syntactic analyzers were established during the shared task of CONLL-X [33], which involved 19 analyzers for 13 different languages. The vast majority of these analyzers were based on treebanks, though each used different machine-learning methods. For example, the NAIST multilingual dependency analyzer includes use of support vector machines and EM algorithms. NAIST dependency analyzer's results are listed in Table 8.2 [49].

English's state-of-the-art supervised syntactic analyzer (Penn Treebank converted to dependencies) corresponds to transition-based systems such as NAIST's [205], graph-based algorithms [128], ensemble parsers [128, 176], and phrase-structure-based analyzers, such as those by Collins et al. [58] and Charniak [47]. The performance of these systems is shown in Table 8.1.

8.1.3   SemiSupervised Dependency Syntax Analysis

Connexor is a semisupervised syntactic dependency analyzer that is commercially available for several languages (English, Spanish, French, German, Swedish, and

Table 8.1 Unlabelled attachment scores of supervised dependency analyzers for English

Analyzer	UAS score
McDonald	93.2
Sagae and Lavie	92.7
Charniak	92.2
Collins	91.7
McDonald and Pereira	91.5
Isozaki et al.	91.4
McDonald et al.	91.0
Yamada and Matsumoto	90.4

Table 8.2 The NAIST supervised dependency syntax analyzer's precision results

Language	Precision
Arab	65.19
Chinese	84.27
Czech	76.24
Danish	81.72
Dutch	71.77
German	84.11
Japanese	89.91
Portuguese	85.07
Slovene	71.42
Spanish	80.46
Swedish	81.08
Turkish	61.22
Bulgarian	86.34

Finnish). This analyzer is based on the functional dependency grammar approach developed by Tapanainen and Järvinen [189] and is composed of three elements: lexicon; morphologic disambiguation module focused on subcategorical information such as person, gender, and number; and a functional dependency grammar (FDG).

Another semisupervised syntactic analyzer is DILUCT [40]: an algorithm that uses heuristics in order to discover relationships between words and co-occurrence statistics learned in an unsupervised way in order to resolve PP attachment issues.

8.1.4 Unsupervised Syntax Analysis

This approach is relatively new, being one of the seminal works of Yuret [209]. Subsequently, several works—such as grammar bigrams [152], top-down generative models [110], contrastive estimation [182], and nonprojective examples [129] —have applied it to certain adjunction phenomena for syntax analysis.

Gorla et al. [95] propose two methods for elaborating upon an unsupervised dependency analysis system; however, to the best of our knowledge, this system is still under development. Our proposal differs completely from theirs.

Cohen et al. [56] use Bayesian parameter estimation for unsupervised dependency analysis, obtaining a precision of adjunction of 59.4% for sentences shorter than 10 words, 45.9% for sentences shorter than 20 words, and 40.5% for all sentences. They use minimum Bayes-Risk classification for English, and the grammar they obtain is a probabilistic grammar that is trained and tested with only PoS tags.

8.2   Grammar Induction Algorithms

Grammar induction algorithms have used several techniques and tools from natural language processing—including categorial grammar (CG), which was first proposed by Ajdukiewicz in 1935. Despite not being designed as a learning paradigm, CG has been used for similar purposes; for example, GraSp [102] is an algorithm that uses CG.

EMILE [73] is another algorithm based on CG; it learns shallow languages in an incremental way and has been applied to natural languages with the assumption that such languages are shallow. By shallow, we mean the property by which, for any constituent type in a language, there are minimal well-supported units for that kind. EMILE aligns whole sentences (trying to isolate minimal units), which are then used to process longer sequences. This method is efficient because its alignment is not recursive, which also means that EMILE offers only a limited treatment of nested and recursive structures. Given its widespread use, we selected the EMILE grammar inductor for the experiments discussed in Sect. 8.3.3.

Searching for string patterns is another technique used for grammar induction algorithms. The idea of this method is to look for repeating patterns that are expected to have a specific function. An algorithm based on this idea is grammatical bigram (GB) [156] which uses dependency grammar formalism to describe the established relationships between pairs of words. If the algorithm finds that the dependents of a head and its order are completely independent, the grammar is simplified. The procedure consists of learning the optimal parameters (probability of dependency for a head) for which statistical measures are used.

Another algorithm based on the technique of looking for repeating patterns is alignment-based learning (ABL) [193], which is the only expectation–maximization algorithm applied to raw text [31]. Along with EMILE, we chose this inductor for the experiments detailed in Sect. 8.3.2.

Other paradigms are used for grammar inductors not covered in this work, such as memory-based learning [72] and evolutive optimization [109]. Roberts and Atwell provide a detailed review of these methods [170].

8.3   Implementation

Several steps are involved in using the grammar inference algorithms previously described to create a dependency analyzer. We describe the process in four stages:

1. Add morphological tags to improve the inductors' performance (Sect. 8.3.1).
2. Complete grammar induction. We used the grammar induction tools ABL (Sect. 8.3.2) and EMILE (Sect. 8.3.3), both of which are open source and freely available from their author's page.[1]

---

[1]ilk.uvt.nl/~menno/research/software/abl and staff.science.uva.nl/~pietera/Emile/.

3. Tune the grammar inductors' parameters. (Sect. 8.4).
4. Convert the grammar inductors' output (a shallow parse) to dependency relationships. We propose using a simple algorithm based on lexical category
   precedence (see Sect. 8.5).

In the following subsections, we discuss each stage in detail. To illustrate our
procedure, we will use the Spanish CAST-3LB corpus [52] as a means of exem-
plification; however, our approach should work for any language that requires only
a PoS tagger as an external resource.

8.3.1   PoS Tagging of Raw Text

CAST-3LB is a Spanish dependency-tagged corpus that we use as a gold standard
for comparison against our annotated version. However, in order to simulate real
situations, our algorithm was given access only to CAST-3LB's raw text.

We generated a PoS-tagged version of the raw CAST-3LB corpus using the TnT
tagger [25], which was trained using the Spanish CLiC-TALP Corpus.[2] The TnT
tagger has a performance of 94% with these settings [142]. Gambino and Calvo
[80] discuss the benefits of adding morphological tag information prior to grammar
induction.

Sample input text follows. The English translation of the following text is "since
then, he entered into a state of complete silence." In addition, b gives the added PoS
information.

(a) desde entonces entró en silencio absoluto.
(b) desde/sps00   entonces/rg   entró/vmis3s0   en/sps00
    silencio/ncms000 absoluto/aq0ms0./Fp

8.3.2   ABL's Output Processing

The output from ABL shows the corresponding (possibly nested) chunks of the
input text. Consider the following sentence: "The reserves of gold are valued at 300
USD per gold troy ounce." The words in bold, below, correspond to the numbers
that are also in bold, thus indicating hypothesis (rule number) 305.

las reservas en oro se valoran en_base_a 300_dólares
estadounidenses por cada **onza troy** de oro.

---

[2]http://clic.fil.ub.es.

@@@(0,1,[3])(15,16,[2])(2,3,[152])(8,9,[154])(0,16,
[0])(13,14,[280])(14,15,[281])(9,10,[285])(4,5,[288])
(3,4,[292])(1,2,[297])(10,11,[302])**(11,13,[305])**(3,8,
[153])(10,13,[<u>291</u>])(9,15,[155])

The numbers 11, 13 show the beginning and end of the chunk. Note that hypothesis 291 (shown underlined) consequently comprises hypotheses 302 and 305 because it covers words from 10 to 13. From this notation, we derive the production rules for each sentence as follows:

```
0  →  [3]  [297]  [152]  [153]  [155]  [2]
    3  →  las
  297  →  reservas
  152  →  en
  153  →  [292] [288] valoran en_base_a 300_dólares [154]
      292  →  oro
      288  →  se
      154  →  estadounidenses
  155  →  [285] [291] [280] [281]
      285  →  por
      291  →  [302] [305]
         302  →  cada
         305  →  onza troy
      280  →  de
      281  →  oro
2  →  .
```

Grouped Chunks

(las reservas en (oro se valoran en_base_a 300_dólares estadounidenses) (por (cada (onza troy)) de oro).)

In some cases, chunks were not added as rules because they were used only once —for example, *valoran*.

8.3.3   EMILE's Output

Emile's output is a set of grammar production rules, as in
The boy plays with the ball.
The boy plays in the park.

```
[0] → the boy[1] | [2]
[1] → plays with the ball
[2] → plays in the park
```

It is also capable of producing grouped chunks, as in
(the boy (plays with the ball) [1] [0])
(the boy (plays in the park) [2] [0])

8.4   Parameter Selection for Grammar Inductors

In order to find the best parameters for grammar inductors, we compared the output
of our inductors with the output obtained using the gold standard CAST-3LB. We
compared the location of opening and closing parentheses. For example, the following sentence ("We cannot remember either why they came.") shows the original
CAST-3LB chunking and a sample output after grammar induction.

CAST-3LB:

(tampoco recordamos ((por qué) llegaron).)

Grammar Inductor:

(tampoco (recordamos (por qué) llegaron.))

The first and third opening parentheses are in the same position as the first and
third closing parentheses (shown in bold). From here, we computed the Recall,
Precision, and F-score measures as follows. Note that these measures were used
only for parameter selection in this middle stage (Table 8.3).

$$\text{Recall} = \frac{\text{\# of coincident parenthesis}}{\text{Total \# of parenthesis in Gold Standard}}$$

Table 8.3 ABL with different parameters test

Corpus	Recall	Precision	F-score
Parameters	Alignment method: **Biased**		
	Selection method: **Branch**		
Raw (%)	17.58	21.19	19.22
Raw + PoS (%)	**17.60**	21.27	19.26
Parameters	Alignment method: **Biased**		
	Selection method: **Leaf**		
Raw (%)	14.27	26.21	18.48
Raw + PoS (%)	14.56	26.63	18.82
Parameters	Alignment method: **Default**		
	Selection method: **Branch**		
Raw (%)	16.88	23.64	**19.69**
Raw + PoS (%)	16.96	23.50	19.70
Parameters	Alignment method: **Default**		
	Selection method: **Leaf**		
Raw (%)	11.69	**31.24**	17.01
Raw + PoS (%)	12.39	**31.24**	17.74

$$\text{Precision} = \frac{\#\text{ of coincident parenthesis}}{\text{Total }\#\text{ of parenthesis in Induction}}$$

F-Score combines recall and precision into one score. We selected $\beta = 1$ so that recall and precision are equally weighed.

$$F_\beta = \frac{\left(\beta^2 + 1\right) * \text{Precision} * \text{Recall}}{\left(\beta^2 * \text{Precision} + \text{Recall}\right)}$$

EMILE provides the following selection of parameters:

1. **total_support_percentage** of context/expression of a particular type.
2. **expression_support_percentage** for an expression in a determined context.
3. **context_support_percentage** of appearances in a context along with expression of certain type.
4. **rule_support_percentage** of characteristic expressions for a type that can be substituted by one of the referred types in the rule. A rule will be incorporated into the grammar only if this percentage is exceeded.

Tables 8.4 and 8.5 show the performances obtained with different parameters. We show the best, the default (in italics), and the worst 4 F-scores—however, note that precision is highest in just one of these cases.

Table 8.4 Parameter selection for EMILE (no PoS Tags, i.e., raw text only)

A	2	3	4	Recall (%)	Prec. (%)	F (%)
50	20	20	25	**9.75**	53.72	**16.51**
60	30	30	30	9.72	54.17	16.49
40	40	40	20	9.71	53.95	16.46
50	40	40	25	9.68	54.39	16.44
75	*50*	*50*	*50*	*9.53*	*55.06*	*16.25*
50	30	30	25	9.47	**54.90**	16.16
70	30	30	35	7.50	42.71	12.76
80	50	50	40	7.47	42.84	12.72
70	50	50	35	7.46	42.91	12.71

Table 8.5 Parameter selection for EMILE (using PoS Tags, i.e., raw + PoS)

1	2	3	4	Recall (%)	Prec. (%)	F (%)
70	70	70	35	9.55	**54.96**	**18.91**
50	20	20	25	**9.80**	53.78	16.57
60	20	20	30	9.69	54.43	16.45
70	20	20	35	9.67	54.58	16.42
70	60	60	35	9.45	54.75	16.11
75	*50*	*50*	*50*	*9.40*	*53.06*	*15.97*
70	30	30	35	7.40	42.61	12.61
70	50	50	35	7.30	42.55	12.46

ABL provides three alignment methods: default, biased, and all. It also provides three selection methods: first, leaf, and branch. Table 8.5 shows the results of testing with different parameters.

8.5   From Chunks to Dependency Relations

The CAST-3LB and output from the grammar inductors can be regarded as chunks of constituents. In this section, we explore a simple mechanism for transforming these constituent chunks into dependencies. First, we review some considerations regarding this conversion.

Ninio [147] points out that the relationship between constituents and dependencies is formally weak. Grammatical relationships are primary to a dependency grammar and, as such, do not have a role within the dependency approach. However, these relationships can be derived from one representation to the other. Marneffe et al. [123] generated typed dependency trees from constituent trees using a constituent grammar analyzer; they later identified the constituent heads following rules proposed by Collins et al. [58].

Robinson [171] points that one important difference between both representations is that the dependency approach uses only terminal categories, while the constituent approach uses categories of a higher degree; despite this, there is a systematic correspondence between the trees produced by each approach. We propose a series of rules by which it is possible to convert one representation into the other.

Gelbukh et al. [88] proposes a procedure based in heuristics and coded as 15 rules in order to mark the head so that a constituent corpus may be converted into a dependency corpus with an accuracy estimated at 95%. Civit et al. [53] propose a similar method based on linguistically motivated rules, which are encoded in a *head table*, but they do not provide an evaluation of that method.

In this work, we look for a simple yet effective way of completing such conversion given that we do not have constituent tags available—only PoS tags.

8.5.1   Lexical Categories Precedence

Hengeveld [101] suggests that there exist common lexical hierarchies among the majority of languages, including both flexible its and inflexible languages, when referring to the linguistic regularity of such languages. He does not, however, mention application to syntactic analysis. On the other hand, Genthial et al. [91] suggest the existence of a lexical category (LC) hierarchy for the construction of syntactic structures. When applying these structures, however, Genthial et al. code them into rules in a similar manner as Gelbukh et al. [88] and Civit et al. [53].

We propose using a LC hierarchy to determine the head for dependency analysis starting from a shallow parse. Our procedure is described in the following pseudo-algorithm.

```
function convert (syntactic groups, head)
```

1. get the most deep-nested syntactic group
2. obtain words and LC from this syntactic group
3. compare the LC of the group
4. mark the word with highest LC precedence as head of this group
5. mark other words as dependent
6. convert(rest of syntactic groups, head of the group)
   end function.

In order to obtain the correct LC hierarchy, we used the original syntactic groups found in the CAST-3LB gold standard. Iteratively, we adjusted the LC hierarchy until a representative sample group of sentences of the gold standard were parsed correctly. The LC hierarchy we obtained is listed below from highest (1) to lowest (12) precedence. The symbols in parentheses correspond to the 3LB tagging system. However, as we will show later, this hierarchy can be easily adapted to a different tagging system.

1. Verb (v)
2. Adverb (rg)
3. Noun (n)
4. Adjective (a)
5. Pronoun (p)
6. Negation (rn)
7. Subordinated conjunction (cs)
8. Preposition (s)
9. Determiner (d)
10. Coordinated conjunction (c)
11. Interjection (i)
12. Punctuation symbols (f)

We found that additional information conveyed by the tags was not necessary for the correct identification of hierarchical position—i.e., vmis3s0 is simplified to v. Additional information (such as person, gender, number, or tense) is discarded for verbs. This simplification is done for every PoS.

For the previously studied example, "since then, he entered into a state of absolute silence," the chunks are as follows:

```
((since/s then/rg) he/p entered/v (in/s (absolute/a
silence/n)./Fp)
```

Fig. 8.1 Dependency tree for
sample sentence

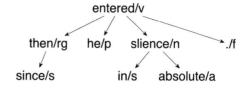

The heads, according to our system are as follows:

Components	HHead selected
since/s then/rg	then/rg
absolute/a silence/n	Silence/n
he/p entered/v	entered/v
then, heads contend at a higher level	
then/rg entered/v	entered/v
silence/n in/s	silence/n
entered/v silence/n	entered/v
entered ./f	entered/v

This yields the tree shown in Fig. 8.1.

8.6   Evaluation of Dependencies

Briscoe et al. [29] suggest evaluating the accuracy of syntactic analyzers based on
the grammar relationships between lemmatized lexical heads. Each tree can be
represented as an *n*-ary representation with *n* triples: each dependency relationship
has a head, dependent, and relationship type. Following this, we compare two
dependency analyses by comparing every triple in them. When we generated extra
triples, they were counted as errors since they are outside the gold standard.

Consider the following example of triple comparison. In this case, the precision
would be 6/7 and recall 6/8.

System output		Gold standard
entered v 0	✓	entered v 0
since s then	✓	since s then
then rg entered	✓	then rg entered
The p entered	✗	–
silence n entered	✓	silence n entered
in s silence	✗	in s entered
absolute a silence	✓	absolute a silence
. f entered	✓	. f entered

Table 8.6 Comparison of our system (DILUCT-GI) with other systems

Measures				Resources			
System	Recall (%)	Precision (%)	F-measure (%)	Dictionaries	Rules	Syntactic annotations	Morphol. annotations
DILUCT	55	47	51	✓	✓	✓	✓
Connexor	42.15	39.60	40.83	✓	✓	✓	✓
DILUCT-GI	31.83	32.36	32.09				✓
TACAT	30	–	–	✓	✓	•	•

CAST-3LB is a Spanish corpus with 3,700 tagged sentences. The best scores we obtained were 31.83% recall, 32.36% precision, and 32.09% F-measure—using ABL, Text, and PoS tags, the Default alignment mode, and Branch selection method. Table 8.6 compares these results with other semisupervised dependency analyzers: DILUCT [40] and TACAT [6]. TACAT is a shallow syntactic parser for Spanish; hence, our results were converted into dependencies for comparison.[3] The compared analyzers use resources such as dictionaries, rules, and syntactic annotations, whereas our proposal uses only morphological annotations; so this comparison might be unfair. However, to the best of our knowledge, there is presently no unsupervised dependency analyzer for Spanish available for comparison.

8.7   Building an English Parser in a Few Days

In order to perform a fair comparison, we need to compare our method with another unsupervised method. For example, for English, the syntactic analyzer from Cohen et al. [56], which uses unsupervised Bayesian parameter estimation, obtains an adjunction precision of 59.4% for sentences shorter than 10 words, 45.9% for sentences shorter than 20 words, and 40.5% for all sentences. The grammar is inferred based on PoS tags with no words, and its output is not a dependency tree. Gorla et al. [95] did not report results of their unsupervised dependency analyzer.

We did an implementation of a parser for English based on the Susanne corpus [178], which consists of 7500 annotated English sentences. As before, we used only the raw text and morphological tags of this corpus, ignoring all syntactic information as input for our syntactic analyzer and used the annotation as the gold standard.

The Susanne corpus is annotated following the Susanne analytic scheme. Genabith et al. [89] recommend converting a corpus to the Xbar notation to minimize the creation of context free grammar (CFG) rules for grammar induction; therefore, we used the Xbar annotated version of the Susanne corpus, which was

---

[3]Results kindly provided by Jordi Atserias, Technical University of Catalonia.

Table 8.7 DILUCT-GI
dependency analyzer results
for the Susanne corpus

Corpus	Recall	Precision	F-score
Parameters	Alignment method: **Biased**		
	Selection method: **Branch**		
Raw + PoS (%)	40.41	40.33	**40.37**
Parameters	Alignment method: **Default**		
	Selection method: **Branch**		
Raw + PoS (%)	39.03	38.97	39.00

created by Nick Cercone.[4] This corpus's morphological tags are different from those of other corpora; for example, pronouns are tagged as nouns and adjectives and adverbs are classified together. Based on the previous LC hierarchy, we obtained, and by retesting our sample sentences, we quickly obtained the Susanne English LC hierarchy, which follows.

- Verb (V)
- Auxiliary verb (have, be, being)
- Auxiliary (Aux)
- Noun (N)
- Adjective_1 (Aeasy)
- Adjective_2 (A)
- Preposition (P)
- Determiner (D)
- Predeterminer (PreDet)
- Conjunction (C)

Results are shown in Table 8.7 for the best parameters found in Sect. 8.4.

8.8   Conclusions and Future Work

Although not directly comparable, our system's performance with English suggests that using a bigger corpus for Spanish may result in better performance—the corpus for English has 7500 sentences, whereas 3LB for Spanish has only 3500.

For dependency analysis, ABL had better performance than EMILE. EMILE stores all content-expression pairs during the induction process in order to create a new nonterminal as part of the grammar; therefore, van Zaanen and Adriaans [193] believe that EMILE will obtain better results with a big corpus (more than 100,000 sentences). On the other hand, ABL uses a greedy algorithm that stores all possible constituents found before selecting the best, which allows ABL to have a better performance with small corpora.

---

[4]Available at www.student.cs.uwaterloo.ca/~cs786s/susanne/.

We obtained better results using a combined corpus of words and tags; the improvement was relatively small but constant in all configurations tested. During the alignment process, the information provided by tags helps to disambiguate constituents that belong to several lexical categories.

We found that the ABL grammar inductor had better performance than that reported by its authors, who tested with the Wall Street Journal corpus in English [193] and obtained a recall rate of 12%. The Biased-Branch configuration for ABL obtained the highest recall (17.60%), while the Default-Leaf configuration obtained the highest precision (31.24%). As expected, the more syntactic groups found, the less precision they have.

We presented a model for dependency analysis, which can be reasonably easy to adapt to other languages, based on unsupervised learning of raw text annotated with morphological tags. To the best of our knowledge, this would be the first unsupervised (after adding PoS tags) dependency analyzer for Spanish; when used for English, our analyzer achieved results that are within the results of the state-of-the-art analyzer. Despite there still being room for improvement, our proposed model alleviates some intrinsic limitations—such as those associated with using grammar inductors for learning [93]—by adding morphologic information before the induction process begins as well as a novel system for converting a shallow parse into a dependency analysis by means of an LC precedence hierarchy. Our method can be used for languages where linguistic resources are scarce, given that morphologic tags are available. We believe that the romance languages (at least) share a similar lexical precedence hierarchy; however, proving this (as well as testing with other corpora) is left for future work.

Chapter 9
Multiple Argument Handling

A sentence can be regarded as a verb with multiple arguments. The plausibility of each argument depends not only on the verb but also on other arguments. Measuring the plausibility of verb arguments is necessary in several tasks, such as semantic role labeling, where grouping verb arguments and measuring the plausibility increases performance [70, 135]. Metaphor recognition also requires knowledge of verb argument plausibility in order to recognize uncommon usages, which would suggest either the presence of a metaphor or a coherence mistake (e.g., *drink the moon in a glass*). Malapropism detection can use the measure of the plausibility of an argument to determine word misuse [24]—such as in *hysteric center* instead of *historic center*, *density has brought me to you* instead of *destiny has brought me to you*, *a tattoo subject* instead of a *taboo subject*, and *don't be ironing* instead of *don't be ironic*. Furthermore, anaphora resolution consists of finding referenced objects, thus requiring (among other things) information about the plausibility of the arguments at hand, i.e., what kind of filler is more likely to satisfy the sentence's constraints. For example, *The boy plays with **it there**, **It** eats grass*, and *I drank **it** in a glass*.

Determining verb argument plausibility can be seen as collecting a large database of semantic frames with detailed categories and examples that fit these categories. For this purpose, recent works take advantage of existing, manually crafted resources such as WordNet, Wikipedia, FrameNet, VerbNet, and PropBank. For example, Reisinger and Paşca [164] annotate existing WordNet concepts with attributes and extend *is-a* relationships based on Latent Dirichlet Allocation on Web documents and Wikipedia, and Yamada et al. [206] explore extracting hyponym relationships from Wikipedia using pattern-based discovery and distributional similarity clustering. The problem with the semantic frames approach for this task is that semantic frames are too general. For example, Korhonen [112] considers the verbs *to fly, to sail,* and *to slide* similar and finds a single subcategorization frame

---

This chapter has been written with Kentaro Inui and Yuji Matsumoto.

© Springer International Publishing AG 2018
A. Gelbukh and H. Calvo, *Automatic Syntactic Analysis Based on Selectional Preferences*, Studies in Computational Intelligence 765,
https://doi.org/10.1007/978-3-319-74054-6_9

for all the three. On the other hand, approaches based on n-grams are too particular, even when used with a very big corpus (such as the Web), have two problems: unavailable combinations and counts that are biased by various syntactic constructions. For example, solving the PP attachment for *extinguish fire with water* using Google[1] yields 319,000 hits for *fire with water* and 32,100 hits for *extinguish with water*, which results in *(*extinguish* (*fire with water*)) instead of (*extinguish* (*fire*) *with water*). Thus, we need a way to smoothen these counts. Since Resnik [166], selectional preferences have been used to do so for verb-to-class preferences, and Agirre and Martinez [1] have used generalization to then transform verb-class-to-noun-class preferences. More recently, McCarthy and Carroll [124] have disambiguated nouns, verbs, and adjectives using automatically acquired selectional preferences as probability distributions over the WordNet noun hyponym hierarchy; they then evaluate with Senseval-2. However, these works have a common problem: they address each argument for a verb separately.

9.1 One Argument Is not Enough

Consider the following sentences: *There is hay at the farm. The cow eats it.*

We would like to connect *it* with *hay*, not with *farm*. From selectional preferences, we know that the object of *eat* should be something edible; so we can say that *hay* is more edible than *farm*, thus solving the issue. We have similar knowledge from semantic frames, but in a broader sense, there is an ingester and an ingestible.

However, this information can be insufficient in cases where the selectional preference depends on other arguments from the clause. For example: *The cow eats hay, but the man will eat it.*

In this case, it is not enough to know that *it* should be edible: the resolution also depends on who is eating. In this case, it is unlikely that *the man* will eat *hay*; so the sentence might refer to the fact that *the man* will eat *the cow*. The same happens with other of verb arguments. For example, the FrameNet peripheral arguments for the ingestion frame include instrument and place. However, some things are ingested with an instrument—e.g., soup is eaten with a spoon, while rice is eaten with a fork or chopsticks (depending on who is eating). Plausible argument extraction allows us to construct a database dictionary for this kind of information, which can be regarded as common sense since it is possible to use large blocks of text to automatically learn the types of activities performed by groups of entities (see Fig. 9.1).

The goal of our work is to construct such a database. To do so, we need to obtain information related to selectional preferences and semantic frames extraction.

---

[1]Google query as of April 2010.

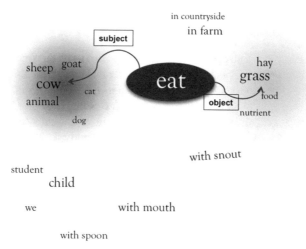

Fig. 9.1 A verb linking groups of related arguments

In the next sections, we will present related work that has been organized into different approaches (Sect. 9.2). We will present a proposal based on the word-space model (Sect. 9.3) as well as a proposal based on language modeling (Sect. 9.4). Then, we will present our major contributions (Sect. 9.5), which consist of interpolated probabilistic Latent Semantic Indexing Model (PLSI) for handling three correlated variables. In each section, we will present several experiments to find how different parameters affect behavior as well as to compare the different approaches.

9.2   Approaches for Learning Verb Argument Preferences

The problem of automatic verb argument plausibility acquisition can be studied from several points of view. From the viewpoint of the type of information extracted, we can find related work for selectional preferences and semantic frames extraction. From the viewpoint of selectional preferences, the task is focused on automatically obtaining classes of arguments for a given verb and syntactic construction. From the viewpoint of semantic frames, arguments are grouped by the semantic role they have, regardless of their syntactic construction; this viewpoint emphasizes the distinction between core (indispensable) and peripheral arguments. On the other hand, we can consider the viewpoint of how information is represented: the task can be regarded as a case of statistic language modeling, where the missing argument should be inferred with high probability by taking in the given context, or it can be regarded as a word-space model task such as that frequently seen in IR systems. In the next sections, we will present work related to this task from each different viewpoint.

9.2.1 Selectional Preferences

Selectional preferences acquisition can be regarded as one of the first attempts to automatically find argument plausibility; earlier attempts dealt with simpler *<verb, argument>* pairs. Since the learning resource is sparse, each work uses a generalization or *smoothing* mechanism to extend coverage. Resnik [166] uses WordNet to generalize the object argument. Agirre and Martinez [1] use a class-to-class model so that both the verb and the object argument are generalized by belonging to a WordNet class. McCarthy and Carrol [126] acquire selectional preferences as probability distributions over the WordNet noun hyponym hierarchy and use argument relationships other than the object-argument. Padó and Lapata [150] combine semantic and syntactic information by using corpora with semantic role annotations (i.e., FrameNet and PropBank) to estimate their model and then apply class-based smoothing using WordNet.

9.2.2 Subcategorization Frames

The following works deal with the problem of semisupervised argument plausibility extraction from the subcategorization frames extraction approach. Salgeiro et al. acquire verb argument structures by generalizing nouns using a named entity recognizer (IdentiFinder) and then using the noisy channel framework for argument prediction. Examples of the kind of information they work with include the following: ***organization** bought **organization** from **organization***, ***thing** bought the outstanding shares on **date***, and (sometimes without generalization) *the cafeteria bought extra plates*.

The work by Kawahara and Kurohashi [107] is also semisupervised; they generalize using a manually created thesaurus. To find case frames, they use the closest argument together with the verb, thus obtaining verb sense disambiguation for cases similar to the example motivating us, which is presented in Sect. 9.1.

9.2.3 Word-Space Model

Traditionally, using information retrieval, words have been represented as documents and semantic context as features so that it is possible to build co-occurrence matrices or word spaces where each intersection of a word and context shows the frequency count of each. This approach has been recently used with syntactic relationships as well [150]. An important issue within this approach is the similarity measure chosen for comparing words (documents), given its features. Popular similarity measures range from simple [such as Euclidean distance, cosine, and

Jaccard's coefficient [117]] to more complex (such as Hindle's and Lin's) measures [119].

In the next sections, we will present a simple proposal within the approach of the word-space model (Sect. 9.3) and then present two algorithms within the language modeling approach (Sect. 9.4).

9.3 A Word-Space Model

We begin with a simple model to explore the possibilities of the last two approaches. Here, we propose a model based on the word-space model. Since we want to consider argument co-relation, the following information is used:

$P(v, r_1, n_1, r_2, n_2)$, where v is a verb, r_1 is the relationship between v and n_1 (noun) as subject, object, preposition, or adverb. r_2 and n_2 are analogous. If we assume that n has a different function when used with another relationship, then we can consider that r and n form a new symbol, called a. Thus, we can simplify our 5-tuple to $P(v, a_1, a_2)$. We want to know, given a verb and an argument a_1, which a_2 is the most plausible; we can write this as $P(a_2|v, a_1)$. For PLSI, this can be estimated using

$$P(a_2, v, a_1) = \text{Sum}(Z_i, P(z) \cdot P(a_2|z) \cdot P(v, a_1|z)).$$

For the word-space model, we can build a matrix where a_2 is the rows (documents) and v, a_1 are features. Since this matrix is very sparse, we use a thesaurus to smooth the argument values. To do so, we loosely followed the approach proposed by McCarthy et al. [125] for finding the predominant sense; however, in this case, we used the k nearest neighbors of each argument a_i to find the prevalence score of an unseen triple given its similarity to all triples present in the corpus, measuring this similarity between the arguments. In other words, just as McCarthy et al. [125] and Tejada et al. [190, 191] did for WSD, we have each similar argument vote for the plausibility of each triple.

$$Prevalence(v, x_1, x_2) = \frac{\displaystyle\sum_{<v,a_1,a_2>\ \in T} sim(a_1, x_1) \cdot sim(a_2, x_2) \cdot P_{MLE}(v, a_1, a_2)}{\displaystyle\sum_{<v,a_1,a_2>\ \in T} sim\_exists(a_1, a_2, x_1, x_2)}$$

where T is the whole set of <*verb*, *argument*$_1$, *argument*$_2$> triples, P_{MLE} is the maximum likelihood of <*verb*, *argument*$_1$, *argument*$_2$>, and

$$sim\_exists(a_1, a_2, x_1, x_2) = \begin{cases} 1 & \text{if } sim(a_1, x_1) \cdot sim(a_2, x_2) > 0 \\ 0 & \text{otherwise} \end{cases}$$

To measure the similarity between the arguments, we built a thesaurus using the method described by Lin [119], using the Minipar browser [119] over

short-distance relationships, i.e., we previously separated subordinate clauses. We
obtained triples $<v, a_1, a_2>$ from this corpus, which were counted and then used to
build the thesaurus and as a source of verb and argument co-occurrences.

9.3.1   Evaluation

We compared these two models in a pseudo-disambiguation task such as that
presented by Weeds and Weir [204]. First, we obtained triples $\langle v, a_1, a_2 \rangle$ from the
corpus. Then, we divided the corpus into two parts: training (80%) and testing
(20%). With the training part, we trained the PLSI model and created the WSM.
This WSM was also used to obtain the similarity measure for every pair of arguments a_2, a'_2. This enables us to calculate *Feasibility* (v, a_1, a_2). For evaluation, we
artificially created 4-tuples $\langle v, a_1, a_2, a'_2 \rangle$, which were formed by taking every triple
$\langle v, a_1, a_2 \rangle$ from the testing corpus and generating an artificial tuple $\langle v, a_1, a'_2 \rangle$ by
choosing a random a'_2 with $r'_2 = r_2$, and making sure that this new random triple
$\langle v, a_1, a'_2 \rangle$ was not present in the training corpus. This task consists of selecting the
correct tuple. Ties occur when both tuples are given the same nonzero score (see
Table 9.1).

We compared two models based on the statistical language model (see
Sect. 9.3.2.1) and the word-space model approaches, respectively. Using the patent
corpus from the NII Test Collection for Information Retrieval System (NTCIR-5)
Patent [78], we parsed 7,300 million tokens and then extracted the chain of

Table 9.1 Pseudo-disambiguation task sample: choose the right option

Verb	Arg	Option 1	Option 2
add	subj: I	obj: gallery	obj: member
calculate	obj: flowrate	subj: worksheet	subj: income
read	obj: question	answer	stir
seem	it	just	unlikely
go	overboard	subj: we	subj: they
write	subj: he	obj: plan	obj: appreciation
see	obj: example	in: case	in: london
become	subj: they	obj: king	obj: park
eat	obj: insect	subj: it	subj: this
do	subj: When	obj: you	obj: dog
get	but	obj: them	obj: function
fix	subj: I	obj: driver	obj: goods
fix	obj: it	firmly	fresh
read	subj: he	obj: time	obj: conclusion
need	obj: help	before	climb
seem	likely	subj: it	subj: act

relationships in a directed way—that is, for the sentence *X add Y to Z by W*, we extracted the triples <add, subj-X, obj-Y>, <add, obj-Y, to-Z>, and <add, to-Z, by-W> and obtained 706 M triples in the form $<v, a_1, a_2>$. We considered only chained asymmetric relationships to avoid false similarities between words that co-occur in the same sentence.

Following Weeds and Weir's method [204], we chose a mix of 20 high- and low-frequency verbs and extracted every triple $<v, a_1, a_2>$ present in the triples corpus for each. Then, we performed experiments with the PLSI and WSM algorithms.

We experimented with different numbers of topics for the latent variable z in PLSI and with different numbers of neighbors from the Lin thesaurus for the WSM expansion. Results are shown in Table 9.2 for individual words: 10 neighbors for WSM and 10 topics for PLSI. Figure 9.2 shows the average results for different neighbors and topics.

9.3.2 Analysis

We have shown results for an algorithm within the WSM approach for unsupervised plausible argument extraction and compared them with those of a traditional PLSI approach, obtaining particular evidence to support that it is possible to achieve

Table 9.2 Precision (P) and Recall (R) for each verb for 10 neighbors (WSM) and 10 topics (PLSI)

Verb	Triples	WSM-10		PLSI-10	
		P	R	P	R
eat	31	0.98	0.92	1.00	0.04
seem	77	0.88	0.09	0.64	0.38
learn	204	0.82	0.10	0.57	0.22
inspect	317	0.84	0.19	0.43	0.12
like	477	0.79	0.13	0.54	0.24
come	1,548	0.69	0.23	0.78	0.17
play	1,634	0.68	0.18	0.69	0.19
go	1,901	0.81	0.25	0.80	0.15
do	2,766	0.80	0.24	0.77	0.19
calculate	4,676	0.91	0.36	0.81	0.13
fix	4,772	0.90	0.41	0.80	0.13
see	4,857	0.76	0.23	0.84	0.20
write	6,574	0.89	0.31	0.82	0.15
read	8,962	0.91	0.36	0.82	0.11
add	15,636	0.94	0.36	0.81	0.10
have	127,989	0.95	0.48	0.89	0.03
average	11,401	0.85	0.30	0.75	0.16

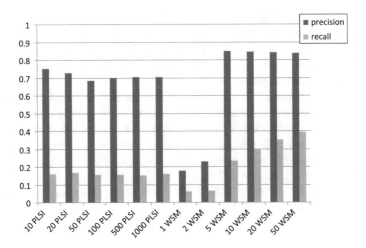

Fig. 9.2 Results for (topics)-PLSI and (neighbors)-WSM

good results with the method that votes for common triples using a distributional thesaurus. The results appear to be consistent with previous works that use distributional thesauruses [37, 190, 191] (see Fig. 9.2): adding information increases coverage with little sacrifice of precision.

We used no other resource after the dependency parser (such as named entity recognizers or labeled data training for machine-learning algorithms) so that from this stage, the algorithm is unsupervised.

To further develop this approach, it is possible to experiment with the upper limit of the increasing coverage since each neighbor from the thesaurus adds noise. We have experimented with building the thesaurus using the same corpus and found significant differences when the encyclopedia corpus was used to build the dictionary since a broader and richer context could be found.

Future work may explore the effect of using other similarity measures, as well as of constructing a similarity table with simpler objects—for instance, a single noun instead of a composite object.

In the next section, we explore other proposals within the language model.

9.3.2.1   Language Modeling

We regard the task of finding the plausibility of a certain argument for a set of sentences as estimating a word given a specific context (see Fig. 9.3). Particularly, for this work, we consider context as the grammar relationships for a particular verb:

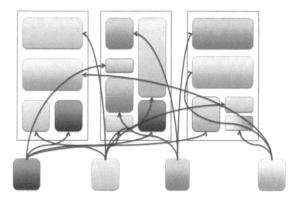

Fig. 9.3 Each document in PLSI is represented as a mixture of topics

$$P(w, c) = P(c) \cdot P(c|w) \tag{9.1}$$

which can be estimated in many ways—particularly, by using a hidden Markov model or using latent variables to smooth for PLSI:

$$P(w, c) = \sum_{z_i} P(z) \cdot P(w|z) \cdot P(c|z)$$

The conditional probability can be calculated from n-gram frequency counts.

9.4 The Dependency Language Model

Most previous work in SLM has been devoted to speech recognition tasks [55, 174] using maximum entropy models. Mostly because of space limitations, these models are usually limited to sequential 3-gram models. However, several works have shown that relying only on sequential n-grams is not always the best strategy [81, 82]. Consider the example borrowed from [81]: [*A baby*] [*in the next seat*] *cried* [*throughout the flight*]. An n-gram model would try to predict *cried* from *next seat*, whereas a dependency language model (DLM) would try to predict *cried* from *baby*.

In this section, we explore creating a DLM to obtain feasible scenario fillers, which can be regarded as extracting selectional preferences [166] but with a broader context for each filler. We show in Sect. 9.4.1.1 how this additional information helps in choosing the best filler candidate. Then, in Sects. 9.4.1.2 and 9.4.1.3, we present our implementations of two models for creating a DLM: one based on PLSI (Sect. 9.4.1.2) and the other based on k-nearest neighbors (KNN) (Sect. 9.4.1.3).

In Sect. 9.4.2, we describe our experiments for comparing both algorithms in a pseudo-disambiguation task. Our results are analyzed in Sect. 9.4.3.

9.4.1 Models for Plausible Argument Estimation

9.4.1.1 Feasible Scenario Fillers

Consider that we want to find the most feasible thing eaten given the verb *to eat*. Since *eat* has several senses, this filler could be food or a material, depending on who is eating (see Tables 9.3 and 9.4). In Table 9.3, the subject is *cow*. Count, in that table, represents the total number of counts that voted for *cow* combinations divided by the number of sources. These tables show our system's (KNN DLM) actual output.

In Table 9.4, the subject is *acid*. It is possible, in that table, to see different adequate fillers depending on the subject doing the action.

If we consider the problem of estimating $P(a_2|v, a_1)$ instead of estimating only $P(a_2|v)$—where a_1 and a_2 are arguments and v is a verb—the data sparseness problem increases. This has been solved mainly by using external resources such as WordNet [1, 126, 166]; semantic-role annotated resources such as FrameNet and PropBank [150]; a named entity recognizer such as IdentiFinder [177]; or other manually created thesauruses [107].

Table 9.3 Feasible arguments for (*eat*, subject: *cow*)

1.3.3	Verb	Argument 1	Argument 2	Count	Sources
1	eat	subj: cow	obj: hay	0.89	2
2	eat	subj: cow	obj: kg/d	0.49	1
3	eat	subj: cow	obj: grass	0.42	12

Table 9.4 Feasible arguments for (*eat*, subject: *acid*)

	Verb	Argument 1	Argument 2	Count	Sources
1	eat	subj:acid	obj:fiber	2	2
2	eat	subj:acid	obj:group	1	4
3	eat	subj:acid	obj:era	0.66	2
4	eat	subj:acid	away	0.25	40
5	eat	subj:acid	obj:digest	0.19	2
6	eat	subj:acid	of:film	0.18	2
7	eat	subj:acid	in:solvent	0.13	1
8	eat	subj:acid	obj:particle	0.11	4
9	eat	subj:acid	obj:layer	0.10	3

In this section, we aim to determine the extent to which information from the corpus itself can be used to estimate $P(a_2|v, a_1)$ without using additional resources. To do so, two techniques are used for dealing with the data sparseness problem; we describe both in the next section.

9.4.1.2 PLSI—Probabilistic Latent Semantic Indexing

As stated before, we can regard the task of finding the plausibility of a certain argument for a set of sentences as estimating a word given a specific context. Since we want to consider argument co-relation, we have

$$P(v, r_1, n_1, r_2, n_2)$$

where v is a verb, r_1 is the relationship between the verb and n_1 (noun) as subject, object, preposition, or adverb. r_2 and n_2 are analogous. If we assume that n has a different function when used with another relationship, then we can consider that r and n form a new symbol, called a. Thus, we can simplify our 5-tuple $P(v, r_1, n_1, r_2, n_2)$ to $P(v, a_1, a_2)$.

We want to know, given a verb and an argument a_1, which a_2 is the most plausible argument, i.e., $P(a_2|v, a_1)$. We can write the probability of finding a particular verb and two of its syntactic relationships as

$$P(v, a_1, a_2) = P(v, a_1) \cdot P(a_2|v, a_1),$$

which can be estimated in several ways. Particularly, for this work, we use PLSI [103] because we can exploit the concept of latent variables to deal with data sparseness.

The probabilistic latent semantic indexing model (PLSI) introduced by [103] arose from latent semantic indexing [69]. The model attempts to associate an unobserved class variable $z \in Z = \{z_1, \ldots, z_k\}$ (in our case, a generalization of the correlation of the co-occurrence of v, a_1, and a_2) and two sets of observables: arguments and verbs + arguments. In terms of a generative model, it can be defined as follows: a v, a_1 pair is selected with probability $P(v, a_1)$, latent class z is then selected with probability $P(z|v, a_1)$, and an argument a_2 is finally selected with probability $P(a_2|z)$. It is possible to use PLSI [103] in this way, which is also expressed as (2).

$$P(v, a_1, a_2) = \sum_z P(z_i) \cdot P(a_2|z_i) \cdot P(v, a_1|z_i) \qquad (9.2)$$

z is a latent variable capturing the correlation between a_2 and the co-occurrence of (v, a_1) simultaneously. Using a single latent variable to correlate three variables may lead to poor PLSI performance; so in the next section, we explore different ways to use latent semantic variables for smoothing.

9.4.1.3 K-Nearest Neighbors Model

This model uses the k nearest neighbors of each argument to find the plausibility of an unseen triple, given its similarity to all triples present in the corpus, and then measuring this similarity between arguments. See Fig. 9.4 for the pseudo-algorithm of this model.

Since the votes are accumulative, triples that have words with many similar words will get more votes.

Common similarity measures range from Euclidean distance, cosine, and Jaccard's coefficient [117] to Hindle's and Lin's respective measures [119]. Weeds and Weir [204] show that the distributional measure with best performance is Lin's; so we used that measure to smooth the co-occurrence space, following the procedure described by Lin [119].

9.4.2 Evaluation

For these experiments, we used the same setting as that presented in Sect. 9.3.1. We artificially created 4–tuples $\langle v, a_1, a_2, a_2' \rangle$, which were formed by taking all triples $\langle v, a_1, a_2 \rangle$ from the testing corpus and then generating an artificial triple $\langle v, a_1, a_2' \rangle$ by choosing a random a_2' with $r_2' = r_2$ and then ensuring that this new random triple $\langle v, a_1, a_2' \rangle$ was not present in the training corpus. Thus, the task consisted of selecting the correct triple.

For evaluation, as in Sect. 9.3.1, we used the patent corpus from the NII Test Collection for Information Retrieval System (NTCIR-5) Patent [78]. We parsed 7,300 million tokens with the MINIPAR parser [119] and then extracted the chain of relationships in a directed way. That is, for the sentence X *add* Y *to* Z *by* W, we extracted the triples $\langle add, subj-X, obj-Y \rangle, \langle add, obj-Y, to-Z \rangle$, and $\langle add, to-Z, by-W \rangle$

We obtained 177 M triples in the form $\langle v, a_1, a_2 \rangle$.

9.4.2.1 Effects of Added Context

To evaluate the impact of adding more information for verb argument prediction, we created a joint minicorpus that consists of 1,000 triples for each of the verbs

```
for each triple <v,a₁,a₂> with observed count c,
    for each argument a₁,a₂
        Find its k most similar words a₁s₁...a₁sk,  a₂s₁...a₂sk
            with similarities s₁s₁, ..., s₁sk and s₂s₁,...,s₂sk.
        Add votes for each new triple <v,a₁si,a₂sj> += c·s₁si·s₂sj
```

Fig. 9.4 Pseudo-algorithm for the K-nearest neighbors DLM algorithm

from the patent corpus: (*add, calculate, come, do, eat, fix, go, have, inspect, learn, like, read, see, seem, write*). We first estimated an argument's plausibility given a verb $P(a_2|v)$ we then used additional information from other arguments $P(a_2|v, a_1)$ in order to compare results for both models.

For completely new words, it is sometimes impossible to estimate an argument's plausibility; thus, in such cases, we measured precision and recall. Precision measures how many attachments were correctly predicted from the covered examples, and recall measures the correctly predicted attachment from the whole test set. Because we are interested in measuring the precision and coverage of these methods, we did not implement any backoff technique.

9.4.3 Analysis

Operating separately on verbs (one mini-corpus per verb) yields better results for PLSI (precision results of above 0.8) but seems not to affect EXPANSOR KNN. For small amounts of context $P(a_2|v)$ PLSI works better than EXPANSOR KNN; for greater amounts $P(a_2|v, a_1)$ EXPANSOR KNN works better.

In general, PLSI prefers a small number of topics, even for a large corpus (the largest corpus used in the experiments had around 20 topics). Recall for EXPANSOR KNN seems to improve steadily when more neighbors are added, though a small amount of precision is lost. Overall, expanding with a few neighbors (1–5) does not appear useful. In particular, as Fig. 9.5 shows, when recall is very low, precision can be either very high or very low because when few cases are solved, performance tends to be random. In general, the recall results seem low

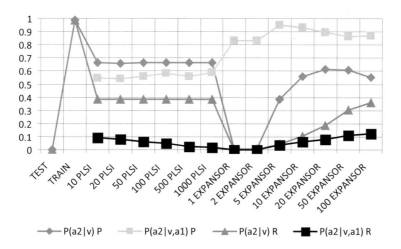

Fig. 9.5 Raw effect of adding more context: prediction based only on the verb versus prediction based on the verb plus one argument. EXPANSOR is the proposed KNN-based model

because we did not use any backoff method. If we compare the precision of the EXPANSOR KNN full model (based on more context), we can think of backing off to PLSI based on pairs $P(a_2|v)$, which would yield the best possible results and which is left as future work.

We evaluated two different dependency language models using a pseudo-disambiguation test. The KNN-based model outperformed the PLSI model when data sparseness was increased by added data. Effective smoothing is achieved by using similarity measures from the L in distributional thesaurus to vote.

Since the PLSI model, we use deals with several arguments with a single latent variable; in the next section, we will present an original improvement that consists of interpolating several PLSI models for handling multiple arguments.

9.5 Interpolated PLSI

In this section, we propose a new model (called interpolated PLSI) that allows multiple latent semantic variables to be used; it is based on the algorithm described in Sect. 9.4.1.2.

9.5.1 iPLSI—Interpolated PLSI

The previous PLSI formula originally used crushes to associate information from a_2 with that from v, a_1 so that one single latent variable results. This causes two problems: data sparseness and correlations between two variables. Thus, we propose a variation that uses interpolation based on each pair of arguments for a triple. We show an interpolated way of estimating the probability of a triple based on the co-occurrences of its different pairs.

Additionally, we test a model that considers additional information. Note that a_i (the latent variable topics) should not be confused with a_1 and a_2 (the arguments).

See Fig. 9.6 for a graphical representation of this concept. Each latent variable is represented by a letter in a small circle. Big circles surround the components of the dependency triple to be estimated. A black dot shows the co-occurrence of two variables; each contributes to the estimation of the triple v, a_1, a_2.

9.5.2 Experiments

We compare these two models in a pseudo-disambiguation task, as shown in Sect. 9.3.1. However, in order to have a wider range of co-occurring words, we

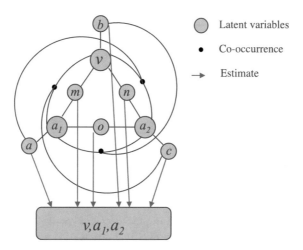

Fig. 9.6 Graphical representation of interpolated probabilistic latent semantic indexing (iPLSI). The tuple v, a_1, a_2 is estimated by using latent variables based on pairs of two variables and/or the pair of a variable and the co-occurrence of two variables

used the UkWaC corpus [75] for these evaluations. The UkWaC corpus is a large balanced corpus of English from UK websites with more than 2 billion tokens.[2] We created two wordsets for the verbs: *play, eat, add, calculate, fix, read, write, have, learn, inspect, like, do, come, go, see, seem, give, take, keep, make, put, send, say, get, walk, run, study, need,* and *become*. These verbs were chosen as a sample of frequently occurring verbs as well as not-so-frequently occurring verbs. They are also verbs that can take a great variety of arguments (i.e., their ambiguity is high), such as *take*. Each wordset contains 1,000 or 2,500 verb dependency triples per verb. The first wordset is evaluated against 5,279 verb dependency triples, while the second wordset is evaluated against 12,677 verb dependency triples, this latter corresponding roughly to 20% of the total number of triples in each wordset (Table 9.5).

9.5.2.1 Results of the Original Algorithm with the New Corpus

Here, we present results for this new corpus using the original PLSI and KNN algorithms. Tests were carried out with one 7-topic variable for PLSI and a 100-nearest-neighbors expansion for KNN. We have shown in Sect. 9.4.2.1 that for estimating the probability of an argument a_2, $P(a_2|v, a_1)$ works better than $P(a_2|v)$. The following table confirms this for different wordset sizes. In most cases, KNN

---

[2]A tool including queries to this corpus can be found at http://sketchengine.co.uk.

Table 9.5 Results of the original PLSI and KNN algorithms for a test with the UKWaC corpus

Mode	Algorithm	Wordset size	Prec.	Recall	F-score
	PLSI	1,000	0.5333	0.2582	0.3479
	KNN	1,000	0.7184	0.5237	0.6058
	PLSI	2,500	0.5456	0.2391	0.3325
	KNN	2,500	0.7499	0.5032	0.6023
	PLSI	1,000	0.4315	0.1044	0.1681
	KNN	1,000	0.8236	0.5492	0.6590
	PLSI	2,500	0.3414	0.0611	0.1036
	KNN	2,500	0.8561	0.6858	0.7615

performs better than the original PLSI in precision and recall (the best of the KNN variations is better than the best of the PLSI variations). Contrary to KNN, PLSI's performance increases with the size of the wordset, probably because there is more confusion in KNN than in PLSI when the same number of topics is used. This can also be seen in Figs. 9.7 and 9.8: recall improves slightly for larger data sets and more topics.

9.5.2.2 Measuring the Learning Rate

This experiment consisted of gradually increasing the number of triples from 125 to 2,000 dependency triples per verb to examine the effects of using smaller corpora. Results are shown in Fig. 9.7. In this figure, KNN outperforms PLSI when more data is added. Indeed, KNN precision is higher in all experiments. The best results for PLSI were obtained using seven topics; the best for KNN were obtained using 200 neighbors.

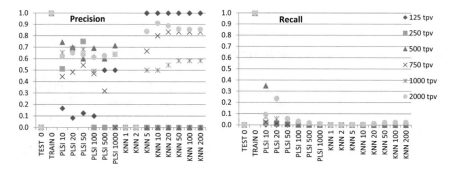

Fig. 9.7 Average of precision and recall for the original PLSI and KNN showing learning rate (each series has different number of triples per verb, *tpv*). No frequency threshold was used. The numbers and the lower part show the number of topics for PLSI and the number of neighbours for KNN

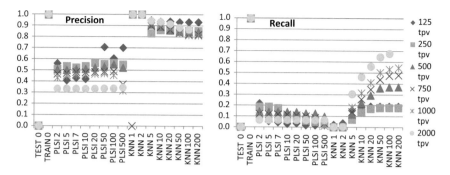

Fig. 9.8 Precision and recall for the original PLSI and KNN with learning rate (each series has different number of triples per verb, tpv). The frequency threshold for triples was set to 4. The numbers and the lower part show the number of topics for PLSI and the number of neighbours for KNN

9.5.2.3   Results with no Pre-filtering

Previous results used a pre-filtering threshold of 4—that is, triples with less than four occurrences were discarded. Here, we present results with no pre-filtering. In Fig. 9.8, the results for KNN fall dramatically. PLSI is able to perform better with 20 topics, which suggests that PLSI is better able to smooth single occurrences of certain triples. KNN is better for work with frequently occurring triples. However, we require a method that can handle occurrences of infrequent words, since pre-filtering implies losing data that could be useful afterward—for example, if *tezgüino* is mentioned only once in the training test, it could be lost when it is needed later. For this reason, we believe it is important to be able to learn information for infrequently mentioned entities. The next section presents results regarding the improvements that are possible when PLSI is used to handle non-filtered items.

9.5.3   iPLSI Results

As presented in Sect. 9.5.1, we tested different models for their effectiveness in combining latent semantic variables. The *mode* part shows the latent variables that were used for these tests. For example, for the *a, c* row, the estimation was carried using Eq. (9.3). Results are presented Table 9.6.

$$P_E(v, a_1 a_2) \approx f_a(v, a_1, a_2) + f_c(v, a_1, a_2) \qquad (9.3)$$

Table 9.6 Comparison of different iPLSI modes, each of which selects a different estimator. KNN is shown in the last row for reference

Mode	Precision	Recall	Mode	Precision	Recall
a, b, c	0.78	0.78	m, n, o	0.83	0.83
a	0.67	0.60	m	0.78	0.77
b	0.44	0.44	n	0.50	0.48
c	0.77	0.77	o	0.84	0.84
a, c	0.62	0.62	m, n	0.77	0.77
a, b	0.78	0.78	m, o	0.83	0.83
b, c	0.76	0.76	n, o	0.84	0.84
KNN	0.74	0.51	a, b, c, m, n, o	0.80	0.80

As Table 9.6 shows, the best results were obtained for o, (using only the information from a_1, a_2) followed by m, o, which combines the information from v, a_1 and a_1, a_2. The m, n, o and n, o modes include n, which has no impact in this test because it is always fixed and which is of little help in deciding the better triple. However, as we show in the following section, in a test with pure n-grams (non-dependency triples, as in all previous tests), the three components (in this case m, n, and o) contribute to the estimation.

9.5.4   The N-Grams Test

We conducted an n-grams test to prove that the three components contribute to the interpolation and avoid the bias that the parser might induce. Trigrams of bigrams were selected from the UkWaC corpus in a similar manner to that discussed for our previous experiments. In this case, however, we did not use dependency relationships; instead, we used sliding windows of hexagrams distributed in trigrams in order to mimic the way function words (e.g., prepositions or determiners) affect triples in the dependency model. The n-grams were extracted for n-grams related to the same verbs described in Sect. 9.5.2. As with the dependency triples task, this task consisted of choosing one of two options. The correct case is always the first triple, though the system does not know this. We used 80% of the trigrams as a base for prediction (training) and 20% for testing. Tests were conducted using the best performance models of the previous experiment (m, n and m, n, o) for anywhere from 500 triples per verb to 5,000 triples per verb.

From Table 9.7, we can see that m, n, o always has the best performance.

Table 9.7 Results of iPLSI for hexagrams grouped as trigrams of bigrams. It shows that it is possible to select the correct trigram amongst two in 75% of the cases

Size, mode	Prec.	Recall	Size, mode	Prec.	Recall
500 m, n	0.75	0.70	2,000 m, n, o	0.77	0.77
500 m, n, o	0.78	0.74	3,000 m, n	0.70	0.70
1000 m, n	0.70	0.70	3,000 m, n, o	0.75	0.75
1000 m, n, o	0.76	0.76	5,000 m, n	0.72	0.72
2000 m, n	0.73	0.72	5,000 m, n, o	0.76	0.76

9.5.5 Analysis

We have seen that the KNN algorithm outperforms single-variable PLSI, and we have studied the learning rate of both algorithms, showing that KNN's recall increases when more data is added, without losing much precision; however, KNN strongly requires a pre-filtering phrase that eventually leads to an important loss of scarcely occurring words. Such words are important for our purposes because filtering them out would prevent us from generalizing rare words in order to measure their plausibility. The interpolated PLSI (iPLSI) algorithm proposed here deals with that issue, yielding better results than single-variable PLSI. We have found that it is possible to select the more feasible hexagram of two options with a 75% recall for raw n-grams grouped as trigrams of bigrams and up to an 83% recall for dependency trigrams. The conducted tests prove that it is possible to select the correct candidate for a triple, which can be regarded as part of a sentence; this allows us to calculate the most plausible argument in a sentence using a broader context as given by a verb and one other argument.

iPLSI outperforms the previous KNN model but still has room for improvement —in particular, estimating the co-occurrence of two arguments simultaneously. The next chapter proposes a model that allows us to do so in order to determine if using more arguments improves argument prediction. Results using that model will then be compared with results from our previous approaches.

Chapter 10
The Need for Full Co-Occurrence

We have previously shown that simultaneously considering three arguments yields better precision than does considering only two, though with a certain loss of recall. Kawahara and Kurohashi [107] differentiate the main verb using the closest argument in order to disambiguate verbs for learning preferences. For example, *play a joke* and *play a guitar* will have different argument preferences; however, in some cases, this is not enough—as can be seen from the following example, where the verb has different meanings depending on a distant argument:

> *Play a scene for friends in the theater* (where *play* means *to act*)
>
> and
>
> *Play a scene for friends in the VCR* (where *play* means *to reproduce*).

Recent works, beginning with Bergsma et al. [13], have proposed a discriminative approach for learning selectional preferences. Ritter et al. [168] and Ó Séaghdha [149] propose a LinkLDA (latent Dirichlet allocation) model with linked-topic hidden variables drawn from the same distribution to model <subject, verb, object> combinations, such as <*man, eats, ramen*> and <*cow, eats, grass*>. However, these works consider, at most, trinary relationships. In order to consider as many arguments as possible for clustering verb preferences, we propose a general model for learning all co-related preferences in a sentence, thus allowing us to measure the plausibility of occurrence. In addition, this model allows us to use statistical as well as manual resources, such as dictionaries or WordNet, to improve predictions. In this work, we give an example that uses PLSI, mutual information, and WordNet in order to measure the plausibility of occurrence.

Furthermore, at this point, there are several particular questions that we seek to answer.

1. With automatic learning, building the co-occurrence table out from real examples can be done in several different ways. Which way is the best?
2. Is joining verb and noun information in a single table better for the model?
3. Can using an SVM that has been trained only with PLSI information outperform the PLSI model itself?

© Springer International Publishing AG 2018
A. Gelbukh and H. Calvo, *Automatic Syntactic Analysis Based on Selectional Preferences*, Studies in Computational Intelligence 765, https://doi.org/10.1007/978-3-319-74054-6_10

4. How does the model perform when the training information varies?
5. Does combining statistical information (from both PLSI and PMI) with manually crafted resource information, such as WordNet, improve results?
6. Should we consider using the model for more than just trinary relationships?

10.1   Method

First, by parsing the UkWaC corpus with MINIPAR [119], thus obtaining a lemmatized dependency representation, we build the resource that counts co-occurrences. The UkWaC corpus [75] is a large balanced corpus of English from UK websites with more than 2 billion tokens. The sentence *Play a scene for friends in the theater* becomes "play obj:scene for:friend in:theater." We then pre-calculate the mutual information statistics between all pairs of words, i.e., (play, obj:scene), (play, for:friend), (play, in:theater), (obj:scene, for:friend), (obj:scene, in:theater), (for:friend, in:theater), and afterward proceed to calculate the topic representation of each word using PLSI.

As discussed previously, in Sect. 9.4.1.2, the probabilistic latent semantic indexing model (PLSI) [103] attempts to associate an unobserved class variable $z \in Z = \{z_1, ..., z_k\}$ with two sets of observable arguments. In terms of generative models, this process can be defined as follows: a document is selected with probability $P(d)$, a latent class z is selected with probability $P(z|d)$, and a word w is selected with probability $P(w|z)$. This can alternatively be represented as Eq. (10.1).

$$P(d, w) = \sum_z P(z_i) \cdot P(d|z_i) \cdot P(w|z_i) \tag{10.1}$$

Given a set of sentences, there are several ways to consider what is a word and what is a document. We can then group documents by either verb or noun. That is, the document *eat* will be for all arguments co-occurring with the verb *eat*, or the document *ball* will be for all other arguments and verbs co-occurring with the noun *ball*—such as *play, with:stripes,* and *for:exercise* (see Table 10.1). On the other hand, documents can only be nouns, and the co-occurrents would then be verbs plus functions (see Table 10.2).

Table 10.1 Co-occurrence table (verbs+nouns)

	With friend	In park
Play	1	1
Eat	1	
Ball		1
Yoyo	1	

Table 10.2 Co-occurrence table (nouns only)

	Play with	Play at
Yoyo	1	
Ball	1	

To summarize, the following different ways of building the PLSI sentence co-occurrence matrix are listed below. fn means function:noun (*with:stripes*), v means verb (*play*), n means noun (*ball*), and vf means verb:function (*play:with*). In some cases, Baroni and Lenci [10] performed experiments with similar matrices. Their nomenclature is indicated in square brackets.

a. (fn|v,fn|v)
bc. (fn,fn), (v,fn) [LCxLC, CxLC]
d. (v|n,fn)
ef. (v,fn), (n,fn) [CxLC,CxLC]
g. (n,vf|nf) [CxCL]
h,i. (n,vf) (n,nf)

Note that modes *a* and *bc* are the same; however, *bc* considers building and training the PLSI model separately for nouns and verbs. The same happens for modes *d*, *ef*, *g* and *hi*. In Sect. 10.2, we detail the results for each of these different settings when building the PLSI sentence co-occurrence matrix.

10.1.1 Assembling SVM Features for Training and Testing

Once the PLSI and PMI resources are built, the training and test sentences are parsed with MINIPAR, but only the first-level shallow parse is used. We mapped features to positions in a vector. Every argument has a fixed offset, i.e., the subject will always be in the first position, the object in the 75th position, arguments beginning with *in* at the 150th position, etc. In this way, the correlation can be captured by an SVM learner. We have chosen a second-degree polynomial kernel that can capture combinations of features. Each argument is decomposed in several subfeatures, which consist of the projection of each word in the PLSI topic space, the pointwise mutual information (PMI) between the target and feature words, and the projection of the feature word in the WordNet space.

The PMI was calculated as

$$PMI(t_1, t_2) = \frac{\log P(t_1, t_2)}{P(t_1, t_2)}$$

Table 10.3 provides an example of the learning data. For the sentence *play a scene for friends in the theater*, the word *scene* is projected in three topics z_1, z_2,

Table 10.3 Simplified example representation for SVM training and testing (one long row)

verb	subj							obj (target)						
play	z_1	z_2	z_3	PMI	wn_1	wn_2	wn_3	z_1	z_2	z_3	PMI	wn_1	wn_2	wn_3
								0.3	0.2	0.5	1	0.8	0.3	0.2
	in							on						
	z_1	z_2	z_3	PMI	wn_1	wn_2	wn_3	z_1	z_2	z_3	PMI	wn_1	wn_2	wn_3
	0.4	0.3	0.8	0.4	0.2	0.4	0.3							
	with							for						
	z_1	z_2	z_3	PMI	wn_1	wn_2	wn_3	z_1	z_2	z_3	PMI	wn_1	wn_2	wn_3
								0.4	0.6	0.4	0.2	0.1	0.9	0.1

and z_3 as 0.3, 0.2, and 0.5, respectively. (note, however, that the experiments considered this projection into 38 topics;) We then calculate the projection in a WordNet space for each word, which is done by calculating the *jcn* distance [104] with regard to the 38 top concepts in WordNet, as shown in Table 10.4. The PMI value for the target word and every target word is also included: (*scene, scene*) 1, (*scene, in theater*) 0.4, and (*scene, for friends*) 0.2.

Table 10.4 Top concepts in WordNet

dry_land_1	money_2
object_1	garment_1
being_1	feeling_1
human_1	change_of_state_1
animal_1	motion_2
flora_1	effect_4
artifact_1	phenomenon_1
instrument_2	activity_1
device_2	act_1
product_2	state_1
writing_4	abstraction_1
construction_4	attribute_1
worker_2	relation_1
creation_3	cognition_1
food_1	unit_6
beverage_1	relationship_3
location_1	time_1
symbol_2	fluid_2
substance_1	

10.2 Experiments

Let us recall the questions we want to answer for these experiments.

1. With automatic learning, building the co-occurrence table out from real examples can be done in several different ways. Which way is the best?
2. Is joining verb and noun information in a single table better for the model?
3. Can using an SVM that has been trained only with PLSI information outperform the PLSI model itself?
4. How does the model perform when the training information varies?
5. Does combining statistical information (from both PLSI and PMI) with manually crafted resource information, such as WordNet, improve results?

Following Weeds and Weir [204], we perform experiments for a pseudo-disambiguation task that consists of changing a target word (in this case, a direct object), at which point the system should identify the most plausible sentence by considering the verb and all of its arguments. For example, for the sentences *I eat **rice** with chopsticks at the cafeteria* and *I eat **bag** with chopsticks at the cafeteria*, the system should be able to identify the first as the most plausible. This experiment setting is similar to experiments previously discussed in Sects. 9.3.1, 9.4.2, and 9.5.2, but in this case, we are considering full phrases instead of only quadruples. We randomly obtained 50 sentences from the WSJ corpus for the verbs: *play, eat, add, calculate, fix, read, write, have, learn, inspect, like, do, come, go, see, seem, give, take, keep, make, put, send, say, get, walk, run, study, need,* and *become*. These verbs were chosen as a sample of both highly frequent and infrequent verbs. They are also verbs that can take a great variety of arguments (i.e., their ambiguity is high), such as *take*. For training, we created wordsets for the same verbs. Each training wordset contains 125, 250, or 500 verb dependency triples per verb; varying this size allows us to answer Question 4. These wordsets were used both to train the PLSI model and also to create the PMI database. The same wordsets were then used to train the SVM. Each sentence was treated as a row, as described in Sect. 10.1.1, with each feature expanded in PLSI subfeatures (topics). We randomly generated two false examples for every good example. For testing, we generated a false example for every existing test example.

At this point, we have not yet included information about WordNet. The first experiment explores different ways of building the co-occurrence matrix, as described in Sect. 10.1 (Questions 1 and 2). We compare using PLSI and PM with and without SVM learning to answer Question 3 (see results in Table 10.5).

Table 10.5 Results of using different modes for building the co-occurrence matrix for PLSI and for using PLSI and PM with and without SVM learning

Train size	Mode	SVM (PLSI and PM)	PLSI * PM	SVM (PLSI and PM)	PLSI * PM	SVM (PLSI and PM)	PLSI * PM
		Coverage	Coverage	Precision	Precision	Recall	Recall
125	a	0.70	0.68	0.58	**0.64**	0.40	0.44
125	bc	0.69	0.68	0.56	0.59	0.39	0.40
125	d	0.83	0.74	0.65	0.59	0.54	0.44
125	ef	0.83	0.74	0.59	0.60	0.48	0.44
125	g	0.86	0.80	0.58	0.56	0.49	0.45
125	hi	0.83	0.72	0.62	0.58	0.51	0.42
250	a	0.78	0.78	0.62	0.59	0.48	0.45
250	bc	0.78	0.77	0.58	0.61	0.45	0.47
250	d	0.88	0.78	0.65	0.60	0.57	0.47
250	ef	0.88	0.78	0.59	0.57	0.52	0.45
250	g	0.90	0.83	0.61	0.55	0.54	0.45
250	hi	0.88	0.79	0.64	0.55	0.56	0.44
500	a	0.86	0.85	0.57	0.54	0.49	0.46
500	bc	0.85	0.85	0.62	0.57	0.53	0.48
500	d	0.92	0.81	**0.68**	0.58	**0.62**	0.47
500	ef	0.92	0.81	0.60	0.49	0.55	0.39
500	g	**0.93**	**0.86**	0.62	0.56	0.58	**0.48**
500	hi	0.92	0.79	0.64	0.56	0.59	0.44

From Table 10.5 and Fig. 10.1, we see that, in all cases, considering all arguments (which is done by adding the SVM learning stage to the PLSI binary co-occurrences) improves performance. In addition, whereas mode g (n,vf|nf) has greater coverage for creating the co-occurrence matrix, mode d (v|n,fn) is always the best choice for precision and recall. We also observe that performance increases with the amount of data in the training wordset.

Both the g and d modes combine verbs and nouns; hence, the answer to Question 2 is yes, it is better to join nouns and verbs in a single table.

10.2.1 Analysis of Adding Manually Crafted Information

In this experiment, we add manually crafted information to the model. As described in Sect. 10.1.1, we add information to the training and testing table regarding the

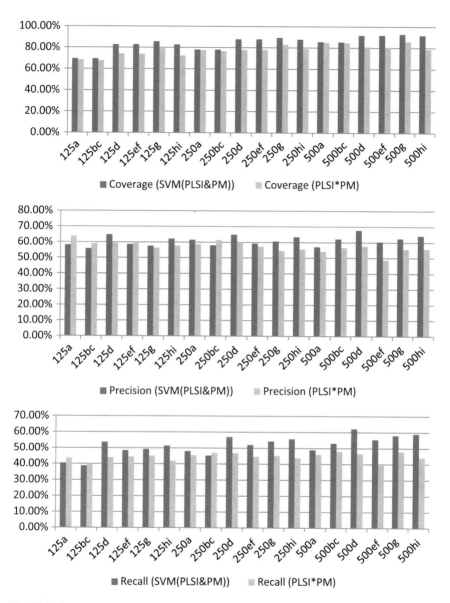

Fig. 10.1 Coverage, precision, and recall results from the first experiment

distance of arguments to the 38 most common concepts in WordNet. Table 10.6 shows the results we obtained.

From Fig. 10.2, we see that, in most cases, combining the three sources of information improves the learning rate; separately, however, PMI provides the highest learning rate. Coverage is always best when the three resources are

Table 10.6 Results of the pseudo-disambiguation task with different settings for PMI, PLSI, and WordNet (WN)

PMI	PLSI	WN	Learning (%)	Coverage (%)	Precision (%)	Recall (%)	F (%)
Wordset 125							
0	0	1	68.36	89.44	54.88	49.09	51.82%
0	1	0	89.59	82.61	66.96	55.23	60.53
0	1	1	92.60	**96.09**	63.23	60.76	61.97
1	0	0	93.63	46.62	**70.98**	33.10	45.15
1	0	1	94.55	94.88	65.85	**62.48**	**64.12**
1	1	0	97.14	83.03	66.09	54.85	59.95
1	1	1	**98.01**	**96.09**	65.26	62.71	63.96
Wordset 250							
0	0	1	67.85	89.49	53.87	48.21	50.88
0	1	0	88.01	87.02	69.44	60.43	64.62
0	1	1	90.82	**96.28**	68.22	**65.69**	**66.93**
1	0	0	93.24	55.18	**70.34**	38.81	50.02
1	0	1	93.78	95.39	64.86	61.87	63.33
1	1	0	96.88	87.12	68.99	60.11	64.24
1	1	1	**97.28**	**96.28**	66.10	64.64	65.36
Wordset 500							
0	0	1	91.09	89.49	46.75	41.84	44.16
0	1	0	86.75	91.58	68.32	62.57	65.32
0	1	1	93.46	96.79	54.37	52.63	53.49
1	0	0	92.95	64.62	65.11	42.07	51.11
1	0	1	93.46	95.72	63.18	60.48	61.80
1	1	0	96.65	91.72	**68.77**	63.08	**65.80**
1	1	1	**96.68**	**97.69**	65.51	**63.41**	64.44
Average							
0	0	1	91.09	89.47	51.83	46.38	48.96
0	1	0	86.75	87.07	68.24	59.41	63.49
0	1	1	93.46	96.39	61.94	59.69	60.80
1	0	0	92.95	55.47	**68.81**	37.99	48.76
1	0	1	93.46	95.33	64.63	61.61	63.08
1	1	0	96.65	87.29	67.95	59.35	63.33
1	1	1	**96.68**	**96.69**	65.62	**63.59**	**64.59**

combined. However, for small amounts of training data, precision is better with PMI only while PLSI gives better support when more data is added. Recall is greater for cases in which WordNet information is used. On average, except for precision, the best values are obtained when the three resources are combined.

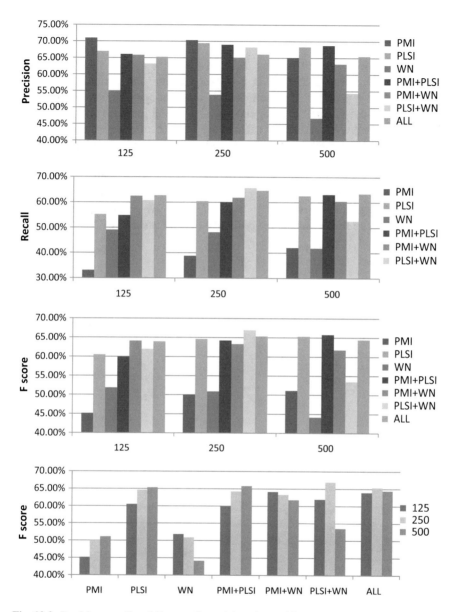

Fig. 10.2 Precision, recall and F score (by training size and by method) for different training corpus' sizes (125, 250 and 500) and feature combinations (PMI, PLSI and WN)

10.3   Conclusions and Future Work

Despite the low amount of training data, we were able to obtain prediction rates above a trivial baseline of random selection between two options. With these experiments, it was possible to determine the impact of using several resources and also to measure the benefit of using an ensemble model for SVM with regard to a simple PLSI model. We found that considering all co-occurrences of an argument in a sentence increases recall by 10%. We also observed that as expected, adding more data increases coverage; however, recall is increased to a greater extent when using SVM over PLSI rather than using PLSI only. Using SVM increases coverage, precision, and recall—even when trained with the same information available to PLSI. This suggests that randomly generating negative examples and applying machine learning to this sample may improve the performance of tasks using topic models.

References

1. Agirre, Eneko, and David Martinez. 2001. Learning Class-to-Class Selectional Preferences. In *Proceedings of the Workshop Computational Natural Language Learning (CoNLL-2001)*, Toulousse, France (A).
2. Agirre, Eneko, and David Martinez. 2002. Integrating Selectional Preferences in WordNet. In *Proceedings of the First International WordNet Conference*, Mysore, India (A).
3. Agirre, Eneko, and David Martínez. 2004. Unsupervised WSD Based on Automatically Retrieved Examples: The Importance of Bias. In *Proceedings of the Conference on Empirical Methods in Natural Language Processing (EMNLP)*, Barcelona, Spain.
4. Apresyan, Yuri D., Igor Boguslavski, Leonid Iomdin, Alexandr Lazurski, Nikolaj Pertsov, Vladimir Sannikov, and Leonid Tsinman. 1989. *Linguistic Support of the ETAP-2 System* (in Russian). Moscow: Nauka.
5. Asoh, H., T. Matsui, J. Fry, F. Asano, and S. Hayamizu. 1999. A Spoken Dialog System for a Mobile Office Robot. In *Proceedings of Eurospeech '99*, 1139–1142. Budapest.
6. Atserias, J., and H. Rodríguez. 1998. TACAT: TAgged Corpus Text Analyzer. *Technical Report LSI-UPC RT-2-98*.
7. Baker, C.F., C.J. Fillmore, and J.B. Lowe. 1998. The Berkeley FrameNet Project. In *Proceedings of the COLING-ACL*, Montreal, Canada.
8. Banerjee, Satanjeev, and Ted Pedersen. 2003. The Design, Implementation, and Use of the Ngram Statistic Package. In *Proceedings of the Fourth International Conference on Intelligent Text Processing and Computational Linguistics*, Mexico City, 370–381.
9. Barros, Flavia de Almeida. 1995. *A Treatment of Anaphora in Portable Natural Language Front Ends to Data Bases*. Ph.D. thesis, University of Essex, UK, 231p.
10. Baroni, M., and A. Lenci, 2009. One distributional memory, many semantic spaces. In *Proceedings of the EACL 2009 Geometrical Models for Natural Language Semantics (GEMS) Workshop*, (pp. 1–8). ACL, East Stroudsburg.
11. Baum, L., T. Petria, G. Soules, and N. Weiss. 1970. A Maximization Technique Occurring in the Statistical Analysis of Probabilistic Functions of Markov Chains. *The Annals of Mathematical Statistics* 41 (1): 164–171.
12. Bechhofer, S., I. Horrocks, P.F. Patel-Schneider, and S. Tessaris. 1999. A Proposal for a Description Logic Interface. In *Proceedings of the International Workshop on Description Logics (DL'99)*, ed. P. Lambrix, A. Borgida, M. Lenzerini, R. Möller, and P. Patel-Schneider, 33–36.
13. Bergsma, S., Lin, D., and R. Goebel, 2008. Discriminative learning of selectional preference for unlabeled text. In: *Proceedings of the 2008 Conference on Empirical Methods in Natural Language Processing*, 59–68.
14. Bogatz, H. The Advanced Reader's Collocation Searcher (ARCS).http://www.elda.fr/catalogue/en/text/M0013.html.

© Springer International Publishing AG 2018
A. Gelbukh and H. Calvo, *Automatic Syntactic Analysis Based on Selectional Preferences*, Studies in Computational Intelligence 765,
https://doi.org/10.1007/978-3-319-74054-6

15. Bolshakov, Igor A. 2004. A Method of Linguistic Steganography Based on Collocationally-Verified Synonymy. In *Information Hiding 2004*, vol. 3200, 180–191. Lecture Notes in Computer Science. Springer (BB).

16. Bolshakov, Igor A. 2004. Getting One's First Million... Collocations. In *Proceedings of the 5th International Conference on Computational Linguistics and Intelligent Text Processing (CICLing-2004)*, vol. 2945, ed. A. Gelbukh, 229–242. LNCS. Springer, 1997 (B).

17. Bolshakov, Igor A. 2004. Two Methods of Synonymous Paraphrasing in Linguistic Steganography (in Russian, abstract in English). In *Proceedings of the International Conference on Dialogue'2004*, Verhnevolzhskij, Russia, June 2004, 62–67.

18. Bolshakov, Igor A., and A. Gelbukh. 2000. A Very Large Database of Collocations and Semantic Links. In *Natural Language Processing and Information Systems. Proceedings of the International Conference on Applications of Natural Language to Information Systems NLDB-2000*, vol. 1959, ed. M. Bouzeghoub et al., 103–114. LNCS. Springer, 2001 (B). www.gelbukh.com/CV/Publications/2000/NLDB-2000-XLex.htm.

19. Bolshakov, Igor A., and A. Gelbukh. 2002. Heuristics-Based Replenishment of Collocation Databases. In *Advances in Natural Language Processing. Proceedings of the International Conference on PorTAL 2002*, Faro, Portugal, vol. 2389, ed. E. Ranchhold, and N. J. Mamede, 25–32, LNAI. Springer.

20. Bolshakov, Igor A., A. Gelbukh. 2004. Synonymous Paraphrasing Using WordNet and Internet. In *Proceedings of the 9th International Conference on Application of Natural Language to Information Systems NLDB-2004*, vol. 3136, ed. F. Meziane, and E. Métais. LNCS. Springer.

21 Bolshakov, Igor A., and Alexander Gelbukh. 1998. Lexical Functions in Spanish. In *Proceedings of the CIC-98, Simposium Internacional de Computación*, Mexico, 383–395. www.gelbukh.com/CV/Publications/1998/CIC-98-Lexical-Functions.htm.

22. Bolshakov, Igor A., and Alexander Gelbukh. 2001. A Large Database of Collocations and Semantic References: Interlingual Applications. *International Journal of Translation* 13 (1–2): 167–187. (B).

23. Bolshakov, Igor A., and Alexander Gelbukh. 2003. On Detection of Malapropisms by Multistage Collocation Testing. In *NLDB-2003, 8th International Conference on Application of Natural Language to Information Systems*, 28–41. Bonner Köllen Verlag (B).

24. Bolshakov, I. A. 2005. An experiment in detection and correction of malapropisms through the web. *LNCS* 3406: 803-815.

25. Brants, Thorsten. 2000. TNT—A Statistical Part-of-Speech Tagger. In *ANLP-2000, 6th Applied NLP Conference*, Seattle, Washington, USA.

26. Bresciani, Paolo, Enrico Franconi, and Sergio Tessaris. 1995. Implementing and Testing Expressive Description Logics: A Preliminary Report. In *Proceedings of the 1995 International Workshop on Description Logics*, Rome, Italy.

27. Brill, Eric. 2003. Processing Natural Language Without Natural Language Processing. In *4th International Conference on Computational Linguistics and Intelligent Text Processing (CICLing 2003)*, ed. Alexander Gelbukh, 360–369. Mexico.

28. Brill, Eric, and Phil Resnik. 1994. A Rule Based Approach to Prepositional Phrase Attachment Disambiguation. In *Proceedings of the Fifteenth International Conference on Computational Linguistics (COLING)* (B).

29. Briscoe, Ted, John Carroll, Jonathan Graham, and Ann Copestake. 2002. Relational Evaluation Schemes. In *Proceedings of the Beyond PARSEVAL Workshop at the 3rd International Conference on Language Resources and Evaluation*, Las Palmas, Gran Canaria, 4–8 (B).

30. Briscoe, Ted, and Nick Waegner. 1993. Generalized Probabilistic LR Parsing of Natural Language (Corpora) With Unification-Based Grammars. *Computational Linguistics* 19: 25–69.

31. Brooks, David J. 2006. Unsupervised Grammar Induction by Distribution and Attachment. In *Proceedings of the 10th Conference on Computational Natural Language Learning (CoNLL-X)*, 117–124. New York City: Association for Computational Linguistics.

32. Burton, R. 1992. Phrase-Structure Grammar. In *Encyclopedia of Artificial Intelligence*, vol. 1, ed. Stuart Shapiro.

33. Buchholz, Sabine, and Erwin Marsi. 2006. CoNLL-X Shared Task on Multilingual Dependency Parsing. In *Proceedings of the Tenth Conference on Computational Natural Language Learning*, 149–164.

34. Calvo, Hiram, and Alexander Gelbukh. 2003. Natural Language Interface Framework for Spatial Object Composition Systems. In *Procesamiento de Lenguaje Natural*, N 31.www. gelbukh.com/CV/Publications/2003/sepln03-2f.pdf.

35. Calvo, Hiram, and Alexander Gelbukh. 2008. Automatic Semantic Role Labeling Using Selectional Preferences with Very Large Corpora. *Computación y Sistemas* 12 (1): 128–150.

36. Calvo, Hiram, and Alexander Gelbukh. 2003. Improving Disambiguation of Prepositional Phrase Attachments Using the Web as Corpus. In *Proceedings of 8th Iberoamerican Congress on Pattern Recognition (CIARP'2003)*, Havana (Cuba), 592–598 (C C).

37. Calvo, Hiram, and Alexander Gelbukh. 2004. Extracting Semantic Categories of Nouns for Syntactic Disambiguation from Human-Oriented Explanatory Dictionaries. In *Computational Linguistics and Intelligent Text Processing (CICLing-2004)*, vol. 2945, ed. A. Gelbukh. Lecture Notes in Computer Science. Springer.

38. Calvo, Hiram, and Alexander Gelbukh. 2004. Unsupervised Learning of Ontology-Linked Selectional Preferences. *In Proceedings of Progress in Pattern Recognition, Speech and Image Analysis (CIARP'2004)*. LNCS. Springer (C).

39. Calvo, Hiram, and Alexander Gelbukh. 2004. Acquiring Selectional Preferences from Untagged Text for Prepositional Phrase Attachment Disambiguation. In *Proceedings of the NLDB-2004*, vol. 3136, 207–216. Lecture Notes in Computer Science (C).

40. Calvo, Hiram, and Alexander Gelbukh. 2006. DILUCT: An Open-Source Spanish Dependency Parser Based on Rules, Heuristics, and Selectional Preferences. In *NLDB 2006*, 164–175.

41. Cano Aguilar, R. 1987. *Estructuras sintácticas transitivas en el español actual*, ed. Gredos. Madrid.

42. Caroli, F., R. Nübel, B. Ripplinger, and J. Schütz. 1994. Transfer in VerbMobil. In *IAI Saarbrücken VerbMobil-Report 11*, May 1994 (C).

43. Carpenter, Bob. 1992. *The Logic of Typed Feature Structures*, vol. 32. Cambridge Tracts in Theoretical Computer Science. Cambridge University Press (C).

44. Carreras, Xavier, Isaac Chao, Lluis Padró, and Muntsa Padró. 2004. FreeLing: An Open-Source Suite of Language Analyzers. In *Proceedings of the 4th International Conference on Language Resources and Evaluation (LREC-04)*, Portugal.

45. Carroll, John, Diana McCarthy (2000) Word sense disambiguation using automatically acquired verbal preferences. In Computers and the Humanities, 34(1–2), Netherlands.

46. Chapman, M., G.I. Davida, and M. Rennhard. 2001. A Practical and Effective Approach to Large-Scale Automated Linguistic Steganography. In *Information Security. Proceedings of International Conference on Information and Communication Security (ICS 2001)*, vol. 2200, ed. G.I. Davida, and Y. Frankel, 156–165. LNCS. Springer.

47. Charniak, Eugene. 2000. A Maximum-Entropy-Inspired Parser. In *NAACL-2000*, 132–139.

48. Charniak, Eugene. 1997. Statistical Techniques for Natural Language Parsing. *AI Magazine* 18: 33–43.

49. Cheng, Yuchang, Masayuki Asahara, and Yuji Matsumoto. 2006. Multi-lingual Dependency Parsing at NAIST. CONLL-X. Nara Institute of Science and Technology.

50. Chomsky, Noam. 1957. *Syntactic Structures*. The Hague: Mouton & Co. (CC).

51. Choueka, Y. 1988. Looking for Needles in a Haystack or Locating Interesting Collocational Expressions in Large Textual Database. In *Proceedings of Conference on User-Oriented Content-Based Text and Image Handling (RIAO'88)*, 609–623.

52. Civit, Montserrat. 2003. *Criterios de etiquetación y desambiguación morfosintática de corpus en español*. Tesis doctoral, Departament de Lingüística, Universitat de Barcelona.

53. Civit, Montserrat, Ma. Antònia Martí, and Núria Bufí. 2006. *From Constituents to Dependencies*, vol. 4139, 141–152. LNCS. Springer.

54. Clark, Stephen, and David Weir. 2002. Class-Based Probability Estimation Using a Semantic Hierarchy. *Computational Linguistics* 28 (2).

55. Clarkson, P. R. and R. Rosenfeld. 1997. Statistical language modeling using the cmu-cambridge toolkit. In *Proceedings of the ESCA Eurospeech*.

56. Cohen, S.B., Kevin Gimpel, and Noah A. Smith 2008. Unsupervised Bayesian Parameter Estimation for Dependency Parsing. In *Advances in NIPS 22*.

57. Collins, Michael, and James Brooks. 1995. Prepositional Phrase Attachment through a Backed-of Model. In *Proceedings of the Third Workshop on Very Large Corpora*, ed. David Yarouwsky, and Kenneth Church, 27–38. Cambridge, Massachussets.

58. Collins, Michael. 1999. *Head-Driven Statistical Models for Natural Language Parsing*. Ph. D. thesis, University of Pennsylvania.

59. Copestake, Ann. 2001. *Implementing Typed Feature Structure Grammars*. The University of Chicago Press.

60. Copestake, Ann, and Dan Flickinger. 2000. An Open-Source Grammar Development Environment and Broad-Coverage English Grammar Using HPSG. In *Second conference on Language Resources and Evaluation (LREC-2000)*, Athens, Greece.

61. Copestake, Ann, Dan Flickinger, and Ivan A. Sag. 1997. *Minimal Recursion Semantics. An Introduction*. CSLI, Stanford University.

62. Craig, J., S. Berezner, C. Homer, and C. Longyear. 1966. DEACON: Direct English Access and Control. In *Proceedings of AFIPS Fall Joint Conference*, San Francisco, CA, vol. 29, 365–380.

63. Cruse, D.A. 1986. *Lexical Semantics*. Cambridge, England: Cambridge University Press.

64. Cuetos, Fernando, Maria Antonia Martí, and Valiña Carreiras. 2000. *Léxico informatizado del Español*. Edicions de la Universitat de Barcelona.

65. Debusmann, Ralph, Denys Duchier, and Geert-Jan M. Kruijff. 2004. Extensible Dependency Grammar: A New Methodology. In *Recent Advances in Dependency Grammar. Proceedings of a Workshop at COLING-2004*, Geneve.

66. Di Eugenio, Barbara. 1993. *Understanding Natural Language Instructions: A Computational Approach to Purpose Clauses*. Ph.D. thesis, University of Pennsylvania, December. Technical Report MS-CIS-93-91.

67. Di Eugenio, Barbara. 1996. Pragmatic Overloading in Natural Language Instructions. *International Journal of Expert Systems* 9.

68. Díaz, Isabel, Lidia Moreno, Inmaculada Fuentes, and Oscar Pastor. 2005. Integrating Natural Language Techniques in OO-Method. In *Computational Linguistics and Intelligent Text Processing (CICLing-2005)*, vol. 3406, ed. Alexander Gelbukh, 560–571. Lecture Notes in Computer Science, Springer.

69. Deerwester, S., S. T. Dumais, G. W. Furnas, Thomas K. L, and Richard Harshman. 1990. Indexing by latent semantic analysis. *Journal of the American Society for Information Science*, 391–407.

70. Deschacht, K. and M. Moens. 2009. Semi-supervised semantic role labeling using the latent words language model. In *Proceedings of the 2009 conference on empirical methods in natural language processing (EMNLP 2009)*, 21–29.

71. Dik, Simon C. 1989. *The Theory of Functional Grammar, Part I: The Structure of the Clause*. Dordrecht: Foris Publications.

72. Domingos, P. 1995. *The RISE 2.0 System: A Case Study in Multistrategy Learning*. Technical Report 95-2. Department of Information and Computer Science, University of California.
73. Dörnenburg, E. 1997. *Extension of the EMILE Algorithm for Inductive Learning of Context-Free Grammars for Natural Languages*. Master's thesis, University of Dortmund.
74. Farreres, X., G. Rigau, H. Rodríguez. 1998. Using WordNet for Building WordNets. In *Proceedings of COLING-ACL Workshop "Usage of WordNet in Natural Language Processing Systems"*, Montreal, Canada.
75. Ferraresi, A., Zanchetta, E., Baroni, M., and S. Bernardini, 2008. Introducing and evaluating ukWaC, a very large web-derived corpus of English. In *Proceedings of the WAC4 Workshop at LREC*, Marrakech, 45–54.
76. Fillmore, Charles. 1968. The Case for Case. In *Universals in Linguistic Theory*, ed. Emmon Bach, and Robert T. Harms, 1–90. Chicago: Holt, Rinehart and Winston (F).
77. Franz, Alexander. 1997. Independence Assumptions Considered Harmful. In *ACL*.
78. Fuji, A., and M. Iwayama, (eds.) 2005. Patent retrieval task (PATENT). In *Fifth NTCIR Workshop Meeting on Evaluation of Information Access Technologies: Information Retrieval*, Question Answering and Cross-Lingual Information Access.
79. Galicia–Haro, Sofia, Alexander Gelbukh, and Igor A. Bolshakov. 2001. Una aproximación para resolución de ambigüedad estructural empleando tres mecanismos diferentes. In *Procesamiento de Lenguaje Natural*, No. 27, Sept 2001. Sociedad Española para el Procesamiento de Lenguaje Natural (SEPLN), Spain, 55–64.
80. Gambino, Omar J., and Hiram Calvo. 2007. *On the Usage of Morphological Tags for Grammar Induction*, vol. 4827, 912–921. Lecture Notes on Artificial Intelligence. Springer.
81. Gao J. and H. Suzuki. 2003. Learning of dependency structure for language modeling. In *Annual Meeting of the ACL archive, Proceedings of the 41st Annual Meeting on Association for Computational Linguistics*, 1: 2003.
82. Gao J., J. Y. Nie, G. Wu, and G. Cao. 2004. Dependence language model for information retrieval. In *Proceedings of the 27th annual international ACM SIGIR conference on Research and development in information retrieval*, 170–177.
83. Gazdar, Gerald. 1982. Phrase Structure Grammar. In *The Nature of Syntactic Representation*, ed. P. Jacobsen, and G.K. Pullum. Boston, Massachussets: Reidel.
84. Gelbukh, Alexander, and Grigori Sidorov. 2002. Automatic Selection of Defining Vocabulary in an Explanatory Dictionary. In *Computational Linguistics and Intelligent Text Processing*, 300–303. Berlin, Heidelberg: Springer.
85. Gelbukh, Alexander, and Grigori Sidorov. 2004. *Procesamiento automático del español con enfoque en recursos léxicos grandes*. IPN.
86. Gelbukh, Alexander, Grigori Sidorov, and Francisco Velásquez. 2003. Análisis morfológico automático del español a través de generación. *Escritos* 28: 9–26.
87. Gelbukh, Alexander, Grigori Sidorov, and Liliana Chanona. 2002. Corpus virtual, virtual: Un diccionario grande de contextos de palabras españolas compilado a través de Internet. In *Proceedings of Multilingual Information Access and Natural Language Processing, International Workshop,IBERAMIA-2002, VII Iberoamerican Conference on Artificial Intelligence*, Spain Seville, ed. Julio Gonzalo, Anselmo Peñas, and Antonio Ferrández, 7–14, 12–15 Nov (G).
88. Gelbukh, Alexander, S. Torres, and H. Calvo. 2005. Transforming a Constituency Treebank into a Dependency Treebank. Submitted to *Procesamiento del Lenguaje Natural*, No. 35, Spain, 1997.
89. van Genabith, Josef, Anette Frank, and Andy Way. 2001. Treebank vs. Xbar-based Automatic F-Structure Anotation. In *Proceedings of the LFG01 Conference*. University of Hong Kong, Hong Kong, CSLI Publications.
90. George, Miller. 1990. WordNet: An On-line Lexical Database. *International Journal of Lexicography* 3 (4): 235–312.

91. Genthial, Damien, Jacques Courtin, and Irene Kowarski. 1990. Contribution of a Category Hierarchy to the Robustness of Syntactic Parsing. In *COLING 1990*, 139–144.

92. Gladki, V. 1985. *Syntax Structures of Natural Language in Automated Dialogue Systems* (in Russian). Moscow: Nauka.

93. Gold, E.M. 1967. Language Identification in the Limit. *Information and Control* 10 (5): 447–474.

94. Goldstein, R.A., and R. Nagel. 1971. 3-D Visual Simulation. *Simulation* 16: 25–31.

95. Gorla, Jagadeesh, Amit Goyal, and Rajeev Sangal. 2007. *Two Approaches for Building an Unsupervised Dependency Parser and their Other Applications*, 1860–1861. AAAI.

96. Grefenstette, G. 1994. *Explorations in Automatic Thesaurus Discovery*. Kluwer.

97. Grosz, B.J., D. Appelt, P. Martin, and F.C.N. Pereira. 1987. TEAM: An Experiment in the Design of Transportable Natural-Language Interfaces. *Artificial Intelligence* 32: 173–243.

98. Haarslev, Volker, and Ralf Möller. 2000. Consistency Testing: The RACE Experience. In *Proceedings of Automated Reasoning with Analytic Tableaux and Related Methods TABLEAUX 2000*, University of St Andrews, Scotland, 4–7 July. Springer.

99. Harris, L. 1984. Experience with INTELLECT: Artificial Intelligence Technology Transfer. *The AI Magazine* 2 (2): 43–50.

100. Hendrix, G.G., E. Sacerdoti, D. Sagalowowicz, and J. Slocum. 1978. Developing a Natural Language Interface to Complex Data. *ACM Transactions on Database Systems* 3 (2): 105–147.

101. Hengeveld, K. 1992. Parts of Speech. In *Layered Structure and Reference in a Functional Perspective*, Benjamins, Amsterdam, ed. M. Fortescue, P. Harder, and L. Kristoffersenpp, 29–56.

102. Henrichsen, P.J. 2002. GraSp: Grammar Learning from Unlabelled Speech Corpora. In *Proceedings of CoNLL-2002*, Taipei, Taiwan, ed. D. Roth, and A. Van den Bosch, 22–28.

103. Hoppe, Th., C. Kindermann, J.J. Quantz, A. Schmiedel, and M. Fischer. 1993 *BACK V5 Tutorial & Manual*. KIT Report 100, Technical University of Berlin.

104. Jiang, J., and D. Conrath, 1997. Semantic similarity based on corpus statistics and lexical taxonomy. In *Proceeding of the International Conference on Research in Computational Linguistics, ROCLING X*.

105. Joshi, Aravind. 1992. Phrase-Structure Grammar. In *Encyclopedia of Artificial Intelligence*, vol. 1, ed. Stuart Shapiro. New York: John Wiley & Sons, Inc. Publishers.

106. Kay, Martin. 1979. Functional Grammar. In *Proceedings of the 5th Annual Meeting of the Berkeley Linguistic Society*, 142–158.

107. Kawahara, D. and S. Kurohashi. 2001. Japanese case frame construction by coupling the verb and its closest case component. In *1st International Conference on Human Language Technology Research*, ACL.

108. Keller, Frank, and Mirella Lapata. 2003. Using the Web to Obtain Frequencies for Unseen Bigrams. *Computational Linguistics* 29: 3.

109. Kirby, S. 2002. Natural Language from Artificial Life. *Artificial Life* 8 (2): 185–215.

110. Klein, D., and C. Manning. 2004. Corpus-Based Induction of Syntactic Structure: Models of Dependency and Constituency. In *Proceedings of the ACL*.

111. Knight, Kevin. 1992. Unification. In *Encyclopedia of Artificial Intelligence*, vol. 2, ed. Stuart Shapiro. New York: John Wiley & Sons, Inc. Publishers.

112. Korhonen, Anna, 2000. Using semantically motivated estimates to help subcategorization acquisi-tion. In *Proceedings of the Joint SIGDAT Conference on Empirical Methods in Natural Language Processing and Very Large Corpora*. Hong Kong. 216–223.

113. Kudo, T., and Y. Matsumoto. 2000. Use of Support Vector Learning for Chunk Identification. In *Proceedings of CoNLL-2000 and LLL-2000*, Lisbon, Portugal.

114. Lucien, Tesnière. 1959. *Eléments de syntaxe structurale*. Paris: Librairie Klincksieck.

115. Lara, Luis Fernando. 1996. *Diccionario del español usual en México*. Digital edition. Colegio de México, Center of Linguistic and Literary Studies.

116. Lázaro Carreter, F. (ed.). 1991. *Diccionario Anaya de la Lengua*, Vox (L).

117. Lee, L., 1999. Measures of distributional similarity. In *Proceedings of 37th ACL.*

118. Li, Hang, and Naoki Abe. 1998. Word Clustering and Disambiguation Based on Co-ocurrence Data. In *Proceedings of COLING '98*, 749–755.

119. Lin, Dekang. 1998. An Information-Theoretic Measure of Similarity. In *Proceedings of ICML'98*, 296–304.

120. Lüdtke, Dirk, and Satoshi Sato. 2003. Fast Base NP Chunking with Decision Trees— Experiments on Different POS Tag Settings. In *Computational Linguistics and Intelligent Text Processing*, ed. A. Gelbukh, 136–147. LNCS. Springer.

121. MacGregor, Robert. 1991. Using a Description Classifier to Enhance Deductive Inference. In *Proceedings of the Seventh IEEE Conference on AI Applications*, Miami, Florida, Feb, 141–147.

122. Manning, C.D., and H. Schütze. 1999. *Foundations of Statistical Natural Language Processing*, 2nd ed. Cambridge, MA, USA: The MIT Press. (M M).

123. de Marneffe, Marie-Catherine, Bill MacCartney, and Christopher D. Manning. 2006. Generating Typed Dependency Parses from Phrase Structure Parses. In *Proceedings of LREC-06.*

124. McCarthy, D. and J. Carroll. 2003. Disambiguating nouns, verbs and adjectives using automatically acquired selectional preferences. *Computational Linguistics* 29(4): 639-654.

125. McCarthy, D., R. Koeling, J. Weeds, and J. Carroll. 2004. Finding predominant senses in untagged text. In *Proceedings 42nd meeting of the ACL*, 280–287.

126. McCarthy, D. and J. Carroll. 2006. Disambiguating nouns, verbs, and adjectives using automatically acquired selectional preferences. *Computational Linguistics* 29(4): 639–654.

127. McDonald, R., K. Lerman, and F. Pereira. 2006. Multilingual Dependency Analysis with a Two-stage Discriminative Parser. In *Proceedings of the CoNLL.*

128. McDonald, R., K. Crammer, and F. Pereira 2005. Online Large-Margin Training of Dependency Parsers. In *Proceedings of the ACL.*

129. McDonald, Ryan, and G. Satta. 2007. On the Complexity of Non-projective Data-Driven Dependency Parsing. In *Proceedings of the IWPT.*

130. McLauchlan, Mark. 2004. Thesauruses for Prepositional Phrase Attachment. In *Proceedings of CoNLL-2004*, Boston, MA, USA, 73–80.

131. Mel'čuk, Igor A. 1996. Lexical Functions: A Tool for the Description of Lexical Relations in the Lexicon. In *Lexical Functions in Lexicography and Natural Language Processing*, ed. L. Wanner, 37–102. Amsterdam/Philadelphia: Benjamins.

132. Mel'čuk, Igor A. 1981. Meaning-Text Models: A Recent Trend in Soviet Linguistics. *Annual Review of Anthropology* 10: 27–62.

133. Mel'čuk, Igor A. 1988. *Dependency Syntax: Theory and Practice*. New York: State University Press.

134. Merlo, Paola, Matthew W. Crocker, and Cathy Berthouzoz. 1997. Attaching Multiple Prepositional Phrases: Generalized Backer-off Estimation. In *Second Conference on Empirical Methods in Natural Language Processing*, ed. Claire Cardie, and Ralph Weischedel, 149–155, Providence, R.I., 1–2 Aug 1997.

135. Merlo, P. and L. Van Der Plas. 2009. Abstraction and generalisation in semantic role labels: propbank, verbnet or both? In *Proceedings of the Joint Conference of the 47th Annual Meeting of the ACL and the 4th International Joint Conference on Natural Language Processing of the AFNLP*, 288–296. Association for Computational Linguistics.

136. Microsoft, Biblioteca de Consulta Microsoft Encarta 2004, Microsoft Corporation. 1994–2004.

137. Minsky, Marvin. 1975. A Framework for Representing Knowledge. In *The Psychology of Computer Vision*, ed. P. Winston, 211–277. New York: McGraw Hill.

138. Mitchell, Brian. 2003. *Prepositional Phrase Attachment Using Machine Learning Algorithms*. Ph.D. thesis, University of Sheffield.

139. Monedero, J., J. González, J. Goñi, C. Iglesias, and A. Nieto. 1995. Obtención automática de marcos de subcategorización verbal a partir de texto etiquetado: el sistema SOAMAS. In *Actas del XI Congreso de la Sociedad Española para el Procesamiento del Lenguaje Natural SEPLN 95,* Bilbao, Spain, 241–254 (M).

140. Montes-y-Gómez, Manuel, Alexander F. Gelbukh, and Aurelio López-López. 2002. Text Mining at Detail Level Using Conceptual Graphs. In *Conceptual Structures: Integration and Interfaces, 10th International Conference on Conceptual Structures (ICCS-2002),* Bulgaria, vol. 2393, ed. Uta Priss et al., 122–136. Lecture Notes in Computer Science. Springer.

141. Montes-y-Gómez, Manuel, Aurelio López-López, and Alexander Gelbukh. 2000. Information Retrieval with Conceptual Graph Matching. In *Proceedings of DEXA-2000, 11th International Conference on Database and Expert Systems Applications,* England, vol. 1873, 312–321. Lecture Notes in Computer Science, Springer.

142. Morales-Carrasco, R., and Alexander Gelbukh. 2003. Evaluation of TnT Tagger for Spanish. In *Proceedings of the Fourth Mexican International Conference on Computer Science ENC'03,* Tlaxcala, México, 18–28.

143. Navarro, Borja, Montserrat Civit, M. Antonia Martí, R. Marcos, and B. Fernández. 2003. Syntactic, Semantic and Pragmatic Annotation in Cast3LB. In *Shallow Processing of Large Corpora (SProLaC), A Workshop of Corpus Linguistics,* Lancaster, UK (N).

144. Nebel, Bernhard. 1999. Frame-Based Systems. In *MIT Encyclopedia of the Cognitive Sciences,* ed. Robert A. Wilson, and Frank Keil, 324–325. Cambridge, MA: MIT Press.

145. Nebel, Bernhard. 2001. Logics for Knowledge Representation. In *International Encyclopedia of the Social and Behavioral Sciences,* Kluwer, Dordrecht, ed. N. J. Smelser, and P. B. Baltes (N).

146. Nebel, Bernhard, and Gert Smolka. 1991. Attributive Description Formalisms.. and the Rest of the World. In *Text Understanding in LILOG,* ed. O. Herzog, and C. Rollinger, 439–452. Berlin: Springer.

147. Ninio, A. 1996. *A Proposal for the Adoption of Dependency Grammar as the Framework for the Study of Language Acquisition,* Volume in Honor of Shlomo Kugelmass, 85–103.

148. *Oxford Collocations Dictionary for Students of English.* Oxford University Press. 2003.

149. Ó Séaghdha, D. 2010. Latent variable models of selectional preference. In: *Proceedings of the 48th Annual Meeting of the Association of Computational Linguistics,* 435–444.

150. Padó, S. and M. Lapata, 2007. Dependency-based construction of semantic space models, *Computational Linguistics* 33(2): 161–199.

151. Pantel, Patrick, and Dekang Lin. 2000. An Unsupervised Approach to Prepositional Phrase Attachment Using Contextually Similar Words. In *Proceedings of Association for Computational Linguistics (ACL-00),* Hong Kong, 101–108.

152. Paskin, M.A. 2001. *Cubic-Time Parsing and Learning Algorithms for Grammatical Bigram Models.* Technical Report, UCB/CSD-01-1148, Computer Science Division, University of California Berkeley.

153. Patel-Schneider, Peter F., Merryll Abrahams, Lori Alperin Resnick, Deborah L. McGuinness, and Alex Borgida. 1996. *NeoClassic Reference Manual: Version 1.0.* Artificial Intelligence Principles Research Department, AT&T Labs Research (P).

154. Pearce, Darren. 2002. A Comparative Evaluation of Collocation Extraction Techniques. In *Proceedings* of the *Third International Conference on Language Resources and Evaluation,* Las Palmas, Canary Islands, Spain.

155. Pereira, F., N. Tishby, and L. Lee. 1993. Distributional Clustering of English Words. In *Proceedings of the 31st Annual Meeting of the Association for Computational Linguistics,* ACL, 183–190.

156. Pereira, Fernando, and Yves Schabes. 1992. Inside-Outside Reestimation from Partially Bracketed Corpora. In *27th Annual Meeting of the Association for Computational Linguistics,* ACL, 128–135.

157. Pineda, L.A., A. Massé, I. Meza, M. Salas, E. Schwarz, E. Uraga, and L. Villaseñor. 2002. *The DIME Project*. Department of Computer Science, IIMAS, UNAM.

158. Pineda, L.A., and G. Garza. 2000. A Model for Multimodal Reference Resolution. *Computational Linguistics* 26 (2): 139–193. (P).

159. Pollard, Carl, and Ivan Sag. 1994. *Head-Driven Phrase Structure Grammar*. Chicago, IL and London, UK: University of Chicago Press.

160. Prescher, D., S. Riezler, and M. Rooth. 2000. Using a Probabilistic Class-Based Lexicon for Lexical Ambiguity Resolution. In *Proceedings of the 18th International Conference on Computational Linguistics*, Saarland University, Saarbrücken, Germany, July–August 2000. ICCL (P).

161. Pullum, Geoffrey K. 1999. Generative Grammar. In *The MIT Encyclopedia of the Cognitive Sciences*, ed. Frank C. Keil, and Robert A. Wilson, 340–343. Cambridge, MA: The MIT Press.

162. Ratnaparkhi Adwait, Jeff Reynar, and Salim Roukos. 1994. A Maximum Entropy Model for Prepositional Phrase Attachment. In *Proceedings of the ARPA Human Language Technology Workshop*, 250–255 (RR).

163. Ratnaparkhi, Adwait. 1998. Statistical Models for Unsupervised Prepositional Phrase Attachment. In *Proceedings of the 36th ACL and 17th COLING*, 1079–1085.

164. Reisinger, J and Marius Paşca. 2009. Latent Variable models of concept-attribute attachment. In *Proceedings 47th Annual Meeting of the ACL and the 4th IJCNLP of the AFNLP*, 620–628.

165. Resnik, Philip. 1993. *Selection and Information: A Class-Based Approach to Lexical Relationships*. Tesis doctoral, University of Pennsylvania (R).

166. Resnik, Philip. 1996. Selectional Constraints: An Information-Theoretic Model and Its Computational Realization. *Cognition* 61 (1–2): 127–159. (R R R).

167. Resnik, Philip. 1997. Selectional Preference and Sense Disambiguation. In *Proceedings of the ACL SIGLEX Workshop on Tagging Text with Lexical Semantics: Why, What, and How?* ACL, 52–57. Washington, DC, USA (R R R R).

168. Ritter, A., Mausam, and O. Etzioni, 2010. A latent dirichlet allocation method for selectional preferences. In: *Proceedings of the 48th Annual Meeting of the Association for Computational Linguistics*, 424–434.

169. Roark, B., and E. Charniak. 1998. Noun-Phrase Co-occurence Statistics for Semi-automatic Semantic Lexicon Construction. In *Proceedings of the 17th International Conference on Computational Linguistics (ICCL)*, 1110–1116. Montréal, Canada: Université de Montréal.

170. Roberts A, and E. Atwell. 2003. The Use of Corpora for Automatic Evaluation of Grammar Inference Systems. In *Proceedings of CL2003: International Conference on Corpus Linguistics, UCREL technical paper number 16*, ed. D. Archer, P. Rayson, A. Wilson, and T. McEnery. UCREL, Lancaster University, 657–661.

171. Robinson, Jane J. 1967. Methods for Obtaining Corresponding Phrase Structure and Dependency Grammars. In *Proceedings of the 1967 conference on Computational linguistics*, 1–25.

172. Rooth, M. 1995. Two-Dimensional Clusters in Grammatical Relations. In *Proceedings of the Symposium on Representation and Acquisition of Lexical Knowledge*, AAAI, Stanford University, Stanford, CA, USA.

173. Roth, D. 1998. Learning to Resolve Natural Language Ambiguities: A Unified Approach. In *Proceedings of AAAI-98*, Madison, Wisconsin, 806–813.

174. Rosenfeld, R., 2000. Two decades of statistical language modeling: where do we go from here?, In *Proceedings of the IEEE* 88(8): 1270–1278.

175. Sag, Ivan A., and Tom Wasow. 1999. *Syntactic Theory: A Formal Introduction*. Center for the Study of Language and Information, CSLI Publications (SSSS).

176. Sagae, K., and A. Lavie. 2006. Parser Combination by Reparsing. In *Proceedings of the HLT/NAACL*.

177. Salgueiro P., T. Alexandre, D. Marcu, and M. Volpe Nunes. 2006. Unsupervised learning of verb argument structures. *Springer LNCS* 3878.

178. Sampson, Geoffrey. 1995. *English for the Computer, The SUSANNE Corpus and Analytic Scheme*. Clarendon Press.

179. Sebastián, N., M.A. Martí, M.F. Carreiras, and F. Cuestos. 2000. *Lexesp, léxico informatizado del español*, Edicions de la Universitat de Barcelona (SS).

180. Shieber, Stuart. 1986. *An Introduction to Unification-Based Approaches to Grammar*. CSLI Publications.

181. Shinyama, Y., Tokunaga, T., and Tanaka, H. 2000. Kairai—Software Robots Understanding Natural Language. In *Third International Workshop on Human-Computer Conversation*, Bellagio, Italy.

182. Smith, N., and J. Eisner. 2005. Guiding Unsupervised Grammar Induction Using Contrastive Estimation. In *Working Notes of the International Joint Conference on Artificial Intelligence Workshop on Grammatical Inference Applications*.

183. Sowa, John F. 1984. *Conceptual Structures: Information Processing in Mind and Machine*. Reading, MA: Addison-Wesley Publishing Co. (S).

184. Sparck-Jones, Karen. 1986. *Synonymy and Semantic Classification*. Edinburgh: Edinburgh University Press.

185. Steele, James (ed.). 1990. *Meaning-Text Theory. Linguistics, Lexicography, and Implications*. Ottawa: University of Ottawa Press.

186. Stetina, Jiri, and Makoto Nagao. 1997. Corpus Based PP Attachment Ambiguity Resolution with a Semantic Dictionary. In *Proceedings of WVLC '97*, 66–80.

187. Suárez, A., and M. Palomar. 2002. A Maximum Entropy-based Word Sense Disambiguation System. In *Proceedings of the 19th International Conference on Computational Linguistics, COLING 2002*, Taipei, Taiwan, vol. 2, ed. Hsin-Hsi Chen, and Chin-Yew Lin, 960–966 (S).

188. Tapanainen, Pasi. 1999. *Parsing in Two Frameworks: Finite-State and Functional Dependency Grammar*. Academic Dissertation. University of Helsinki, Language Technology, Department of General Linguistics, Faculty of Arts.

189. Tapanainen, Pasi, and Timo Järvinen. 1997. A Non-projective Dependency Parser. In *Proceedings of the 5th Conference on Applied Natural Language Processing*, Washington, D.C., 64–71.

190. Tejada, J., Gelbukh A., Calvo, H. 2008a. An innovative two-stage wsd unsupervised method. *SEPLN Journal* 40.

191. Tejada, J., Gelbukh A., Calvo, H. 2008b. Unsupervised WSD with a dynamic thesaurus. In *11th International Conference on Text, Speech and Dialogue. TSD 2008*, Brno, Czech Republic, 8–12 Sept.

192. Vandeghinste, Vincent. 2002. Resolving PP Attachment Ambiguities Using the WWW. In *The Thirteenth Meeting of Computational Linguistics in The Netherlands, CLIN 2002* Abstracts, Groningen, 2002.

193. van Zaanen, M. 2002. *Bootstrapping Structure into Language: Alignment-Based Learning*. Ph.D. thesis, School of Computing, University of Leeds.

194. Volk, Martin. 2000. Scaling Up. Using the WWW to Resolve PP Attachment Ambiguities. In *Proceedings of Konvens 2000*, Ilmenau, Oct 2000.

195. Volk, Martin. 2001. Exploiting the WWW as a Corpus to Resolve PP Attachment Ambiguities. In *Proceeding of Corpus Linguistics 2001*. Lancaster (V V).

196. Watt, W. 1968. Habitability. *American Documentation* 19: 338–351.

197. Webber, Bonnie. 1995. Instructing Animated Agents: Viewing Language in Behavioral Terms. In *Proceedings of the International Conference on Cooperative Multi-modal Communications*, Eindhoven, Netherlands.

198. Weeds, Julie. 2003. *Measures and Applications of Lexical Distributional Similarity*. Ph.D. thesis, University of Sussex.

199. Weinreich, Uriel. 1972. *Explorations in Semantic Theory*. The Hague: Mouton.

200. Weischedel, R.M. 1989. A Hybrid Approach to Representation in the JANUS Natural Language Processor. In *Proceedings of the 27th ACL*, Vancouver, British Columbia, 193–202.

201. Winograd, Terry. 1972. *Understanding Natural Language*. New York: Academic Press. (W).

202. Winograd, Terry. 1983. *Language as a Cognitive Process. Volume I: Syntax*. Stanford University. Addison-Wesley Publishing Company (W).

203. Woods, W.A., R.M. Kaplan, and B.L. Nash-Webber. 1972. *The Lunar Science Natural Language Information System: Final Report*. BBN Report No. 2378. Bolt, Beranek and Newman Inc. Cambridge, MA.

204. Weeds, J. and D. Weir. 2003. A general framework for distributional similarity, In *Proceedings conference on EMNLP* 10: 81–88.

205. Yamada, Hiroyasu, and Yuji Matsumoto. 2003. Statistical Dependency Analysis with Support Vector Machines. In *Proceedings of the 8th International Workshop on Parsing Technologies (IWPT)*, 195–206.

206. Yamada I., K. Torisawa, J. Kazama, K. Kuroda, M. Murata, S. de Saeger, F. Bond and A. Sumida. 2009. Hypernym discovery based on distributional similarity and hierarchical structures. In *Proceedings 2009 Conference on Empirical Methods in Natural Language Processing*, 929–937.

207. Yarowsky, D. 2000. Hierarchical Decision Lists for Word Sense Disambiguation. *Computers and the Humanities* 34 (2): 179–186.

208. Yarowsky, David, S. Cucerzan, R. Florian, C. Schafer, and R. Wicentowski. 2001. The Johns Hopkins SENSEVAL-2 System Description. In *The Proceedings of SENSEVAL-2: Second International Workshop on Evaluating Word Sense Disambiguation Systems*, Toulouse, France, ed. Preiss, and Yarowsky, 163–166.

209. Yuret, Deniz. 1998. Discovery of Linguistic Relations Using Lexical Attraction. Ph.D. thesis, MIT (Y).

210. Zavrel, Jakub, and Walter Daelemans. 1997. Memory-Based Learning: Using Similarity for Smoothing. In *Proceedings of the ACL'97*.

Printed in the United States
By Bookmasters